Edo-Tokyo, Challenging the Urban Fabric

Ueno   Hongo   Yanaka   Nezu   Shitaya

Published in Japan on April 30, 2019 (e-book) / September 30, 2019 (paperback)
by Shokokusha Publishing Co., Ltd.
8-21 Tomihisa-cho, Shinjuku-ku, Tokyo 162-0067 Japan
Tel +81-3- 3359-3231
http://www.shokokusha.co.jp/

Authors: Hosei University Research Center for Edo-Tokyo Studies +
Department of Architecture, Faculty of Engineering and Design, Hosei University +
SCI-Arc + Politecnico di Torino
Publisher: Masanori Shimoide

© 2019 Hosei University Research Center for Edo-Tokyo Studies +
Department of Architecture, Faculty of Engineering and Design, Hosei University +
SCI-Arc + Politecnico di Torino
ISBN978-4-395-32139-1   C3052

Any unauthorized duplication (copying), reproduction, or recording to magnetic, optical, or
other media of the content of this book, whether in whole or in part, is strictly prohibited.
Please contact the publisher for authorization.

謝辞

横山泰子
（法政大学江戸東京研究センター長）

　2018年7月、法政大学江戸東京研究センターは、「FCLT（Future City Laboratory Tokyo、都市東京の近未来）」研究活動の一環としてワークショップ「都市の文脈に挑戦する」を開催しました。法政大学デザイン工学部建築学科、南カリフォルニア建築大学（SCI-Arc）、トリノ工科大学（PoliTo）の学生が東京の近未来について議論し、デザインの提案を行ったのです。学生たちは、上野、本郷、谷中、根津、下谷を実際に歩いて調査し、都市組織を読み、構造を解析するという行為によって江戸東京に挑戦しました。彼らの挑戦的な経験を参加者のみにとどめるのではなく、多くの方々に共有していただきたいと考え、本書を刊行する運びとなりました。ワークショップの開催にあたり、3大学の皆様をはじめ多くの方のお力をいただきましたが、書物の出版にはさらに多くの方々のご尽力を得ました。ここに、すべての関係者の方々、そして本書を手にとってくださった方に感謝の意を表します。

Acknowledgments

Yasuko Yokoyama
(Director, Hosei University Research Center for Edo-Tokyo Studies)

In July 2018, Hosei University Research Center for Edo-Tokyo Studies held the workshop, themed "Challenging the Urban Fabric," as part of the research of the Future City Laboratory Tokyo. Students from the Southern California Institute of Architecture (SCI-Arc) , Politecnico di Torino, and the Faculty of Engineering and Design, Hosei University discussed the future of Tokyo and made design proposals. Based on their fieldwork in Ueno, Hongo, Yanaka, Nezu, and Shitaya, the students read the urban fabric and analyzed its components, thereby grappling with Edo-Tokyo as the subject of their study. This booklet was produced to share this endeavor with a wider community. A large number of individuals from the three universities and beyond contributed toward the holding of the workshop, and so did many others for this publication. Our deepest gratitude is extended to every one of them as well as those reading this booklet.

謝辞　横山泰子　3

FCLT江戸東京国際ワークショップ
「都市の文脈に挑戦する」開催に寄せて　赤松佳珠子　岩佐明彦　6

"知"の向上に向けて　カルラ・バルトロッシ　クリスティアーナ・ロシニョーロ　8

都市の多様性に触れて　ヘルナン・ディアス・アロンソ　10

「都市組織」とは何か　渡辺真理　12

江戸東京の都市組織

東京の都市組織を読む　陣内秀信　20

上野公園を中心とした都市組織

上野公園を中心とした周辺エリア／6つの対象サイト　栗生はるか　46

都市の時代変遷　52

都市の構成要素　54

　暗渠｜根津・千駄木　56　　崖線｜谷中　60　　あんがわ｜下谷　64

　すきま｜東上野　68　　つぶつぶ｜上野 アメ横　72　　かさね｜弥生　76

都市組織に挑戦する

緑を探して――東京の都市構造　クラウディア・カッサテッラ　82

東京の見えない空間　ニコラ・ルッシ　86

伝統と革新の相対的関係　マウロ・ヴォルピアノ　90

レクチャー①　トリノ再構築――都市の未来、ハイブリッド空間　マルコ・サンタンジェロ　94

レクチャー②　LA――対極的な部分が織り成す全体　ジョン・N・ポーン　102

レクチャー③　「江戸東京」という都市のコンセプト　北山 恒　110

新たな都市国家――都会／田舎　ヘルナン・ディアス・アロンソ　118

東京の現状と今後への期待　陣内秀信　122

目　次

Acknowledgments Yasuko Yokoyama 3

Reflecting on "Challenging the Urban Fabric," an International Workshop on Edo-Tokyo Organized by FCLT Kazuko Akamatsu and Akihiko Iwasa 7

Toward Cultivating a Deeper Knowledge Carla Bartolozzi and Cristiana Rossignolo 9

Learning from the Experience of Urban Diversity Hernan Diaz Alonso 11

What is Urban Fabric? Makoto Shin Watanabe 13

The Urban Fabric of Edo-Tokyo

Reading the Urban Fabric of Tokyo Hidenobu Jinnai 21

The Urban Fabric of the Ueno Park Area

The Ueno Park Area: Six Target Sites Haruka Kuryu 47

Chronological Transition of the City 52

Urban Elements 54

buried river | NEZU & SENDAGI 56 cliff edge | YANAKA 60 hard-edged block | SHITAYA 64

gap | HIGASHI UENO 68 grains | UENO AMEYOKO 72 layering | YAYOI 76

Challenging the Urban Fabric

Searching for Green Patterns within Tokyo's Urban Fabric Claudia Cassatella 83

Invisible Spaces in Tokyo Nicola Russi 87

A Challenging Dialectic between Tradition and Innovation Mauro Volpiano 91

Lecture① Reframing Torino: Hybrid Spaces of the Future City Marco Santangelo 95

Lecture② The Idiosynthetic in Los Angeles John N. Bohn 103

Lecture③ The Urban Concept of Edo-Tokyo Koh Kitayama 111

New City States: Urban and Rural Hernan Diaz Alonso 119

The City of Edo-Tokyo and My Hope for its Future Hidenobu Jinnai 123

Contents

FCLT江戸東京国際ワークショップ
「都市の文脈に挑戦する」
開催に寄せて

赤松佳珠子
（法政大学デザイン工学部建築学科主任）

岩佐明彦
（法政大学大学院デザイン工学研究科建築学専攻専攻主任）

2018年度の法政大学江戸東京研究センター「FCLT（Future City Laboratory Tokyo、都市東京の近未来）」プロジェクトの一環として、南カリフォルニア建築大学（SCI-Arc）とトリノ工科大学（PoliTo）を招き、3カ国の学生たちが一緒になって東京とその近未来について思考し、議論を通して最終的にはデザインの提案を行うワークショップを開催することができました。

それぞれがまったく違ったバックボーンを持つ建築学科の学生として、江戸からの歴史を基層構造として持ち、強くその影響を受け継いでいる上野－谷中・根津・千駄木エリアを実際に歩いて調査を行うことで、どのようなことを感じ取り、考え、その魅力と現状の問題点を抽出し、そこから新たな都市の可能性を見出すことができるか。それぞれが受けている建築教育のカリキュラムは違った特徴を持つため、当然違ったアプローチ方法になりますが、現地のサーベイを重ね、作業しながら、多くの議論を通してひとつの提案にまとめていく過程は

学生たちにとってだけではなく、われわれ教員としてもたいへん刺激的であり、新しい発見に満ちた日々となりました。

建築学科主任としてこのようなワークショップが行われたことは、これからの法政大学のみならず、日本の建築学科における教育プログラムとしてたいへん有意義であったと喜ばしく思っており、継続的な開催を望むばかりです。参加された各大学の教員および学生たちにも感謝の意を表したいと思います。

Reflecting on "Challenging the Urban Fabric," an International Workshop on Edo-Tokyo Organized by FCLT

Kazuko Akamatsu
(Head of the Department of Architecture, the Faculty of Engineering and Design, Hosei University)

Akihiko Iwasa
(Head of the Graduate School of Architecture, the Faculty of Engineering and Design, Hosei University)

Inviting students and faculty from the Southern California Institute of Architecture (SCI-Arc) and Politecnico di Torino, Hosei University held a workshop where students from the three universities engaged in active discussion on the city of Tokyo and its future to conceptualize design proposals. The workshop, which was part of the 2018 project of the Future City Laboratory Tokyo (FCLT) at Hosei University's Center for Edo-Tokyo Studies, offered field experience in Ueno, Yanaka, Nezu, and Sendagi – neighborhoods that were developed on the foundations of the urban structure retaining historical continuity and influence from the Edo era.

Through this fieldwork, what can the student participants with each school's unique backbone perceive and think about? What do they identify as attractions and problems within these neighborhoods? Based on the answers to these questions, what urban possibilities can they find? These were the questions I had prior to the workshop. From the angle reflecting each school's unique curricula, the participants studied these neighborhoods in groups. As the students exchanged ideas vigorously and shaped them into proposals – an intellectually stimulating process for the students and faculty alike – the workshop unfolded into days of discoveries.

As the Head of Hosei University's School of Architecture, it is a great pleasure to have held the workshop, a program for architectural instruction that holds significance for the future of Hosei University as well as architectural schools in Japan. As we hope to continue this workshop, we would like to extend our sincere gratitude to all the participants.

"知"の向上に向けて

カルラ・バルトロッシ
（トリノ工科大学建築学研究科長）

クリスティアーナ・ロシニョーロ
（トリノ工科大学都市計画・デザイン学研究科長）

　トリノ工科大学（PoliTo）、法政大学FCLT（Future City Laboratory Tokyo、都市東京の近未来）および南カリフォルニア建築大学（SCI-Arc）の3大学で、「都市の文脈に挑戦する」と題した国際ワークショップを2018年7月東京にて共催することができ、たいへん嬉しく思います。

　法政大学による課題設定はまさに"挑戦"が求められる内容であり、本学の研究・教育の特色でもある学際的アプローチから建築学研究科および都市計画・デザイン学研究科でタッグを組んで挑みました。トリノ工科大学からは、建築・都市設計、都市・景観計画、地理および都市史を専門とする教授陣が、建築学／都市計画学専攻の大学院生を率いて参加し、その知見を共有しました。

　法政大学と本学は、国際的に高い評価を受けている他大学との協力を促進し、学生に有意義な学びの場を与える目的のもと国際化戦略を進めています。多文化共生を誇る国際色豊かな本学は、120カ国以上から5000人を超える留学生を迎え入れており、2018年度QS世界大学ランキングにおいては、上位50位（建築／ビルト・エンバイロメントの部）に入っています。また463本の国際協定を礎に、英語での授業やデュアル・ディグリーを含む多様な教育機会を提供しています。法政大学との国際協定も、教育・学生交流ならびにワークショップの共催を通じて両校の学びに貢献しています。本ワークショップの開催は、研究方法やアイデアの共有を促進し、互いの"知"の向上に資するものです。

　新たな発想へと導く対話を紡ぐこの取り組みは、3大学の関係各位によるご尽力の賜物であり、ここに感謝の意を表すとともに、協力関係のさらなる深化に期待しています。

Toward Cultivating a Deeper Knowledge

Carla Bartolozzi
(Head of the School of Architecture, Politecnico di Torino)

Cristiana Rossignolo
(Head of the School of Planning and Design, Politecnico di Torino)

Politecnico di Torino was glad to collaborate with Hosei University FCLT and the Southern California Institute of Architecture (SCI-Arc) in promoting and organizing the international on-site workshop "Challenging the Urban Fabric" (Tokyo, July 2018).

The topic and the case study proposed by Hosei University was "challenging" in itself, and called for a joint effort by our School of Architecture and School of Planning and Design, in line with the transdisciplinary approach which characterizes our teaching and research activity. In particular, the team of professors from Politecnico di Torino, leading its students from post-graduate architecture and planning courses, brought expertise in the fields of architectural and urban design, urban and landscape planning, geography, and urban history.

Both Hosei and Politecnico di Torino have an on-going internationalization strategy that aims at increasing collaboration with top global universities and at providing fruitful opportunities to students. The Politecnico, a top-fifty university in 2018 QS World University Rankings (Architecture / Built Environment), is a truly international and multicultural university which is currently chosen by over 5,000 international students from more than 120 countries, offering courses in English, double degrees, and a variety of opportunities, thanks to 463 international agreements.

The agreement with Hosei University has already proven to be fruitful, thanks to teaching and student exchanges, and joint workshops. This urban studies workshop further strengthens mutual knowledge and the sharing of methodologies and ideas. We would like to express our gratitude to our partners, and all of our commitment in keeping on our inspiring conversation. We look forward to further collaboration.

都市の多様性に触れて

ヘルナン・ディアス・アロンソ
（南カリフォルニア建築大学ディレクター兼CEO）

　南カリフォルニア建築大学（SCI-Arc）は、日本の奥深い歴史から学ぶとともに現代日本の建築分野の理論的活動に参加するために、長年にわたり教授陣と学生を日本来訪の旅に送り出しており、法政大学とは東京およびロサンゼルスを拠点に10年近くたいへん有意義な交流を続けています。SCI-Arcが築き上げてきた貴重な協力関係の一環として、慶應義塾大学SFC、東京大学、京都精華大学をはじめとする他大学の学生や教授陣とも協働してきました。また、デザイン分野における専門領域の枠を超えて日米間、都市間、大学間の交流を促すべく、東京でのシンポジウム開催も行いました。イタリアのトリノ工科大学（PoliTo）とは初のコラボレーションとなった3大学共催のFCLT（Future City Laboratory Tokyo、都市東京の近未来）国際ワークショップは、地球をぐるっと囲む3大陸からのメンバーが集結した、またとない機会となりました。

　本ワークショップでは、対象を東京都内の上野公園の周囲に設定しました。上野には多様な都市の条件が備わっており、学生チームにとってはこの多様性に触れ、それを分析し表現する良い機会となりました。法政大学の教授陣および教育補助員の方々によって選定された都市の要素には、寺社と住宅が混在する馴染みやすく縁ある地区や大きな公園も含まれ、都市における多様性に着目することができました。各大学、学生にはそれぞれ特徴的な傾向があり、固有の関心を持っていますが、3大学混成チームで取り組んだグループ・ワークの成果は、各対象地区の特徴と各チームのユニークな洞察を反映する結果となりました。

　末筆ながら、今回もまた素晴らしい教育の場を設けてくださった法政大学にSCI-Arcより感謝申し上げるとともに、本ワークショップが、建築的に重要な都市である東京を舞台に、世界各地の学生および教授と協働する機会へと一層の飛躍を遂げることを願っています。

Learning from the Experience of Urban Diversity

Hernan Diaz Alonso
(SCI-Arc Director / CEO)

SCI-Arc has a long tradition of sending students and faculty to Japan to learn from its rich history and participate in its active contemporary architectural discourse. For almost 10 years, SCI-Arc and Hosei University have participated in productive exchange between our faculty and students in Tokyo and Los Angeles. As part of this important, on-going relationship, SCI-Arc has also worked with students and faculty in Japan including those from Keio SFC, University of Tokyo, and Kyoto Seika University. Additionally, SCI-Arc has hosted symposia in Tokyo to intensify the exchange between our countries, cities and schools across multiple disciplines of design. This international workshop with Hosei University and PoliTo was our first Tokyo workshop with a new collaborator from Italy and we were grateful for the opportunity to participate in such an exchange that spans the globe.

The 2018 FCLT (Future City Laboratory Tokyo) International Workshop was an opportunity for three institutions from three countries to come together and study a specific urban area in Tokyo. Ueno, Tokyo has a diverse set of urban conditions that provided student work groups from Hosei University, PoliTo and SCI-Arc opportunities to experience, analyze and articulate a diverse set of urban conditions. Ranging from a large-scale park to an intimate historic residential and religious neighborhood, these varied urban conditions were thoughtfully curated by Hosei faculty and student assistants. This provided the three participating institutions a rich, focused subject for their study in Tokyo. While each institution and student inevitably have their own interests and proclivities, these student groups, made up of members of each institution, were able to quickly assemble, organize and develop studies that not only reflected the specifics of their subject area in Ueno but also illustrate the unique insights of their student group.

We hope that this workshop leads to more opportunities to work with students and faculty from all over the world in the important architectural city of Tokyo and SCI-Arc continues to be grateful to Hosei University for hosting another wonderful event.

「都市組織」とは何か

渡辺真理
（法政大学デザイン工学部建築学科教授）

この書物は法政大学江戸東京研究センターの4つの研究プロジェクト・チームのひとつである「FCLT（Future City Laboratory Tokyo、都市東京の近未来）」チームが2018年7月に実施した国際ワークショップがベースになっている。国際ワークショップを実施するにあたり、イタリア・トリノに拠点を置くトリノ工科大学（以下PoliToと表記）とアメリカ・ロサンゼルスの南カリフォルニア建築大学（以下SCI-Arcと表記）と連携した。PoliToと法政大学は2016年に「学術一般協定」を締結し、2017年度9月にPoliToで行われた国際サマースクール「Beyond the University Dorm」には非公式ながら本学からも教員と学生有志が参加した。SCI-Arcと本学は長い関係を持っている。両校は2012年に「学術一般協定」を締結し、その後毎年、SCI-Arcは東京で、法政大学はロサンゼルスでそれぞれの学舎を使用して教育プログラムを実施している。

2018年春にPoliToのマルコ・サンタンジェロとクラウディア・カッサテッラが来日した際にワークショップのテーマを確定するための会合が持たれたが、そのなかで

「都市組織 tessuto urbano, urban fabric」をテーマとすること、東京のなかでも、江戸－明治－大正－昭和という時代と文化の連続性を感じさせる上野の森の周辺地域を共同調査の対象にすること、また最終発表会ではトリノ、ロサンゼルス、東京の都市組織についての論考を各大学から発表することが定められた。

ところで「都市組織」とは何なのだろうか？　人類は古来、自らの居所である都市とその周囲の地域（テリトーリオ）を記述する努力を続けている。それが都市図や地形図と呼ばれるものなのだが、その表現は時代や国によりさまざまである。fig.1は江戸東京研究センターの機関誌の表紙であるが、江戸東京の中心部を4つの方法で記述したものである。ここには19世紀前半の江戸期古地図、19世紀後半明治期の参謀本部作成の測量図、21世紀の現況を示す航空写真と「図－地」図という4種類の記述がスライスされ並置されている。ここにリプレゼント（代理表象）されているものが紛れもなく「都市組織」であるとして、街路と地所－建物が地形のなかで織り成す

What is Urban Fabric?

Makoto Shin Watanabe
(Professor, Department of Architecture, Faculty of Engineering and Design, Hosei University)

In July 2018, in collaboration with Politecnico di Torino (PoliTo) based in Turin, Italy and the Southern California Institute of Architecture (SCI-Arc) based in Los Angeles, USA, the Future City Laboratory Tokyo team – one of the four research project teams at the Hosei University Research Center for Edo-Tokyo Studies – conducted an international workshop.

Hosei University's 2016 general academic agreement with PoliTo provided the foundation for Hosei University students and faculty to participate in PoliTo's international summer school program, "Beyond the University Dorm," held in September 2017 as their extracurricular activity. In addition, on the basis of Hosei University's 2012 general agreement with SCI-Arc, we strengthened our partnership through hosting annual educational exchanges on campus in Tokyo and Los Angeles. This workshop brought the three universities together for the first time.

In spring 2018, we selected "urban

fabric" (*tessuto urbano*) as the focus of the workshop at a meeting when Marco Santangelo and Claudia Cassatella visited Japan (though Claudia was unable to attend due to a schedule conflict).

So, what is urban fabric? From ancient times, humans have endeavored to describe the cities they inhabit and their surrounding areas (*territorio*). This has produced urban and topographic maps, with their expression varying among nations and periods.

Fig.1 depicts central Tokyo in four ways. It juxtaposes four sections of maps of Tokyo: a historical map of Edo in the former half of the 19th century, a Meiji Period ordnance map in the latter half of the 19th century, an aerial map of the 21st century, and today's solid-void map. Represented in these maps is unmistakably urban fabric, one that reveals a remarkable variety of patterns created by streets, lands, and buildings, reflecting the city's natural environment, climate, culture, economy, land ownership

パターンは、それぞれの都市の自然環境、気候、また文化、経済、土地所有などの制度、建物の建築材料および建築構法などを反映して驚くほど多様である。建築家も都市計画家も都市研究者もそれぞれの立場から都市組織の読解に専念してきた。PoliTo、SCI-Arc、法政大学の3校の教員と学生が、それぞれ慣れ親しんだトリノ（欧州）、ロサンゼルス（米国）、東京（アジア）という3つの都市の都市組織を念頭に置きつつ、上野－谷中－根津という、現在の東京のなかでも「江戸東京」の連続性が垣間見える地域を3校合同で調査研究の対象としたときに、いったいどのような知見が得られるのだろうか？

　本書は3部から構成されている。
　最初の「江戸東京の都市組織」は陣内秀信（法政大学江戸東京研究センター特任教授）とPoliToのクラウディア・カッサテッラ、ニコラ・ルッシ、マウロ・ヴォルピアノの3名が江戸東京の都市組織を論じた記録である。
　ワークショップの初日に陣内秀信が参加した3大学の学生と教員のために「東京の

都市組織を読む」というタイトルで講演を行った。この講演は、参加者が東京の都市組織の特徴とそのなかでの上野、谷中、根津の位置づけを理解するのに非常に有用だった。本書では講演を陣内自らがリライトし文章化してくれたので、このエリアの都市組織を理解するための絶好のイントロダクションになった。
　ワークショップ初日には、酷暑のなか、対象となる6エリアを参加者全員で踏破した。PoliToの3名からの寄稿、「緑を探して──東京の都市構造」（カッサテッラ）、「東京の見えない空間」（ルッシ）、「伝統と革新の相対的関係」（ヴォルピアノ）は、そういった実際の都市体験を踏まえたうえで、彼らが自国のパラダイムと異なる東京とその都市組織をどのように位置づけようとしているかという観点から読み解くと、ことのほか興味深い。

　次の「上野公園を中心とした都市組織」は、ワークショップで取り上げた上野の森周辺の対象エリアの都市組織とその特徴を明らかにするものである。
　「上野公園を中心とした周辺エリア／6つ

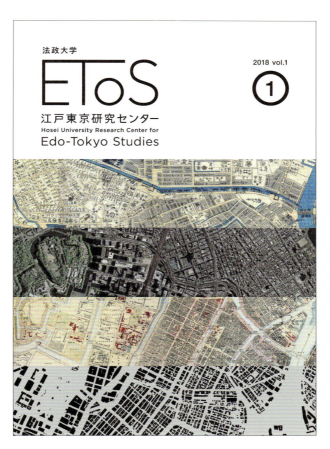

fig.1 法政大学江戸東京研究センター冊子 (2018) 表紙／
The front-page of the 2018 brochure of the Hosei University Research Center for Edo-Tokyo Studies

の対象サイト」という冒頭の文章では栗生はるか（法政大学）が選定された6つの敷地とその特徴について書いている。6つのサイトには、各々の都市空間の特徴が「暗渠」「崖線」「あんがわ」「すきま」「つぶつぶ」「かさね」という短い語句で示されているが、この6つの語句は江戸東京の都市空間に共通する特徴的なものでもある。以下に6つのサイトを列挙しておく。

● 根津・千駄木 Nezu & Sendagi──暗渠 buried river
● 谷中 Yanaka──崖線 cliff edge
● 下谷 Shitaya──あんがわ hard-edged block
● 東上野 Higashi Ueno──すきま gap
● 上野 アメ横 Ueno Ameyoko──つぶつぶ grains
● 弥生 Yayoi──かさね layering

　最後の「都市組織に挑戦する」はワークショップ最終日の発表会に調査した6サイトの学生たちの発表の後で参加した3大学の教員から行われたものである。
　まず、PoliToのマルコ・サンタンジェロ

から「トリノ再構築──都市の未来、ハイブリッド空間」の発表があった。ヨーロッパの代表的な都市のひとつであるトリノの都市組織の変遷を通して、ヨーロッパ都市に特徴的な強固な都市組織について理解することができる。次のSCI-Arcジョン・N・ボーンからの「LA──対極的な部分が織り成す全体」では、レイナー・バンハムの『ロサンゼルス──4つの生態系』(1971)を引用しつつ、すきまだらけのロサンゼルスの都市空間は都市組織でなく空間移動体験に特徴づけられるのではないかと述べている。ここでいう空間移動は無論、自動車によるものである。最後の北山恒の発表、「『江戸東京』という都市のコンセプト」ではヨーロッパの都市を連続壁体都市と呼び、東京はそれに対して「細粒都市」であるという。細粒都市ではソリッド（密）でなくヴォイド（疎）、つまり地割りや街路パターンから都市構造を読み取れるのではないかと北山は論じている。最後に、3人の発表への、SCI-Arcディレクター兼CEOのヘルナン・ディアス・アロンソと陣内秀信からのコメンタリーも掲載した。

and other schemes, construction material, or construction methods. Architects, urban planners, and urbanologists have been devoted to reading urban fabric from their respective standpoints.

What knowledge can the students and faculty from the three universities gain when studying the urban fabric of the Ueno, Yanaka, and Nezu districts – which offer glimpses of connectivity from Edo (Edo-period Tokyo) to current Tokyo – keeping in mind the city they are familiar with, Turin (in Europe), Los Angeles (in America), and Tokyo (in Asia), respectively? This question underlay the motif behind choosing the workshop theme "Challenging the Urban Fabric."

The six-day workshop began with fieldwork on the first day, Monday, and finished with presentations on the final day, Saturday. Despite this rapid pace, typical to international workshops, the students' work scored a satisfactory outcome. This success owes much to the selection of six target neighborhoods and the collection of map information, historical material, and other data by six teaching assistants prior to the workshop. The appeal of the urbanism, diversity, and uniqueness of the six neighborhoods around Ueno Park was instrumental to organizers' and participants' dedication to the workshop activities while informal preliminary presentations by the three universities' faculty members contributed to the substance of the final presentations.

江戸東京の都市組織
The Urban Fabric of Edo - Tokyo

東京の都市組織を読む

陣内秀信
（法政大学江戸東京研究センター特任教授）

はじめに

　文化的な背景の異なる3つの国の大学の建築スクールが協働して、歴史を残す日本らしい文京区、台東区のさまざまな場所を対象とし、リサーチに基づく今後への提案を行うという今回の企画は、東京をより知りたい、より魅力的な都市にしたいと考えるわれわれにとって、じつに大きな意義をもつ。その作業に着手する最初の段階として、東京、特に対象となるエリアの地形・自然条件、歴史的・文化的な背景、そして建築・都市の空間構造の特質について述べてみたい。

1. 東京の都市構造の特徴 ——自然と対話する都市

　まず、東京の原型ともいえる江戸に遡って、この東京の都市空間の特徴を読み解くことから始めよう。東京は、世界の都市の

なかでも珍しく、歴史的に市壁をもたず、都市と田園の境界が明確に区分されず、柔軟に拡大成長ができた一方、都市内にも自然の要素が大いに入り込むという独特の性格をもってきた。小さな中世の城下町を核としながら、1600年頃、徳川家康により江戸幕府が置かれ、江戸城を中心とする巨大な城下町として建設された。その江戸の特徴ある都市空間が、現代東京の中心部の骨格の下敷きをなしているということができる。

　江戸東京においては、都市のグランドデザインは、地形・自然条件に依存していたといえる。1800年代初めに描かれた鳥瞰図を見ると、西洋都市のような人工的なモニュメントは目立たず、逆に大地としての山や丘、森や林、谷と川、濠や堀が織りなす変化に富んだ地形がまずは描かれ、その上に家並み、寺社など建造物がはめ込まれている。西洋都市、中国都市にしばしば見

Reading the Urban Fabric of Tokyo

Hidenobu Jinnai

(Professor, Hosei University Research Center for Edo-Tokyo Studies)

Introduction

This workshop, attended by members of three architectural schools with respective cultural backgrounds, aims to propose urban planning ideas through researching areas that retain historical traces in the Bunkyo and Taito wards of Tokyo. The workshop is of great value as we hope to learn more and increase the appeal of Tokyo as a city. To begin our study, allow me to introduce the target areas' natural and topographic conditions, cultural and historical backgrounds, and architectural and urban spatial structures.

1. Urban Structure of Tokyo
A City Responding to Nature

Let us first read the spatial features of Tokyo by revisiting the city of Edo, which can be described as prototypical Tokyo. Unlike other cities of the world, Tokyo historically has neither city walls, nor clear borders between urban and pastoral lands, thus allowing the city's flexible expansion and incorporation of abundant natural elements. Around the year 1600, the Edo Government was founded by Shogun Ieyasu Tokugawa. Placing Edo Castle at its center, the city of Edo was built as a large castle-town, developed on a small, preexisting medieval castle-town. In this way, the urban space of Edo formed a foundation of the framework of today's central Tokyo.

The grand design of Edo-Tokyo depended on natural and topographic conditions. The bird's-eye view of Edo drawn in the early 1800s depicts not so much the artificial monuments found in Western cities but

fig.1 鍬形蕙斎《江戸名所の絵》(19世紀初頭)／Keisai Kuwagata, *Edo meisho no e* [A picture of the famous places of Edo] (early 19th c.)

られるような強い軸線、幾何学的な構成は見られない[fig.1]。

　この都市は、地形条件が多様で、しかも大きな土木工事も行いながら、独自の構造をつくりあげた。まず、武蔵野台地の突端に江戸城を構え、その南・西・北へ広がる山の手は、凸凹地形を巧みに生かした緑あふれる「田園都市」を生み出し、主に武家屋敷がつくられた。だが、谷あいや低地には商人・職人の町も分布した。

　一方、東の低地では、海の埋め立てを進めながら、運河網が巡るグリッドの計画的な市街地を形成した。舟が行き交い、流通、商業が活発な「水の都」であり、主役は商人・職人だった。この異なる性格・価値を有する都市空間を併せもつという点が東京の魅力の源であり、また強みである。しかも、都市の中心が複数存在し、ひとつの原理に縛られることのない柔軟な構造であるため、時代の変化に多様に対応でき、都市のエネルギッシュな発展を生み出すことができた。

歴史とアイデンティティの継承法

　東京の都心部は、1923年の関東大震災、1945年の戦災の爆撃を受けたため、残っている古い建物が欧米都市に比べ著しく少ない。しかし、歴史的につくられた都市の骨格、道路のネットワーク、水路、都市組織、土地利用の形態をはじめ、江戸を受け継ぐ東京らしい空間的特徴は至るところに見出せる。

　歴史の継承のメカニズムが西洋都市と根本的に異なる。都市にとっての歴史とは何か、空間のアイデンティティとは何かを東京は世界に向けて問題提起する。

　都市のグランドデザインが地形・自然条件に依存していた東京だけに、地形と道路網の関係がまず興味深い。山の手を見ると、広域をつなぐ主要な道路はすべて尾根道を通っており、逆にローカルコミュニティにとって重要な谷道が存在する。そしてその両者を結ぶ坂道が数多くある[fig.2]。

　武蔵野台地の東端に、江戸城が置かれた。

fig.2 道路網と地形の関係／Relationship between road network and topography

the versatile topography consisting of mountains, hills, forests and groves, and moats and canals. Added on top of this are housing rows, temples, shrines, and other buildings. There are no pronounced axes and geometric forms unlike some Western and Chinese cities [fig.1].

The city of Edo, boasting this versatile topography, shaped a unique structure with large-scale civil engineering works. Edo Castle sat at the end of the Musashino plateau, and the "high city" (*yamanote*) – spreading toward the south, west, and north of the castle – formed a garden city dominated by samurai residences that took advantage of hilly landforms. In the valleys and lowlands, merchant and artisan towns developed.

In the eastern lowland, a planned grid town with a canal network was formed, reclaiming land from the sea. This lowland facilitated a vibrant "city of water," a trade and transportation hub for merchants and artisans where ships frequently passed through.

These spatial features and values created both the appeal and advantage of Tokyo as a multi-center city operating beyond one set of urban planning principles. As such, Tokyo has been able to advance itself energetically by adapting to the changing times.

Continuity of History and Identity

Destroyed by the Great Kanto Earthquake of 1923 and wartime bombings in 1945, central Tokyo retains considerably fewer old buildings than typical Western cities. Yet, the historical framework of a city – the road network, waterways, urban tissues, land use, and other spatial features continuing from the city of Edo – are prevalent.

 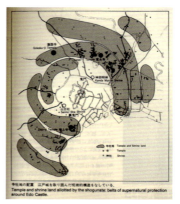

fig.3　都市形態の連続性／Continuity of urban morphology

fig.4　寺社地の配置／Arrangement of temple and shrine districts

　それを受け継ぐ皇居の存在は東京の都市空間にとってきわめて重要な役割をもつ。地形を生かし、また大胆に改造を加えて、濠の水と緑で包まれた有機的な形の城が生まれた。このような江戸城＝皇居の在り方が、東京全体の有機的な成り立ちをまさに象徴しているように見える。

　銀座、丸の内の上空からの写真と江戸時代の地図を比べると、パリやローマのような古い都市組織を切断する都市の大改造がなかった様子がわかる。都市の基本構造はおおむね受け継がれているといえる [fig.3]。

　西洋都市のように城壁がなかったことが、東京の大きな特徴だといえる。その代わり、都市の周縁部、エッジに地形を生かして寺町をつくる傾向があり、それが都市の拡大とともにバウムクーヘンのように、いくつかの円環を形成してきた [fig.4]。

土地と建物

　西洋都市と日本の都市の大きな違いは、不動産の在り方の違いとなって現れる。それは地図を比較するとよくわかる。有名なジャンバティスタ・ノリのローマ地図（18世紀）と江戸切絵図（19世紀半ば）とを比べると違いは歴然としている [fig.5]。ノリの地図では、建物部分と空地部分が明確に区別され、都市のヴォイド（広場、街路、中庭など）が明確に浮かび上がる。一方、切絵図は敷地の境界線しかなく、建物は一切描かれていない。どこが空地かはわからない。ハードな情報は少ないのに対し、土地利用の分類（武家地、町人地、寺社地）、屋敷ごとに家族名、町人地の名前、寺の名前が書き込まれ、ソフト情報が豊富なのである。

　イタリアの都市では、不動産として価値があるのはもっぱら建物であり、不動産地図において建物ごとに番号がふられ、それに対応する台帳には、建物のフロアごとに所有者、面積、用途が記載されている。いかに建物に価値があり、それが継続されて

fig.5 ノリによるローマの地図（18世紀）と江戸の尾張屋版切絵図（19世紀中頃）／Nolli's map of Rome (18th c.) ; Owariya Seishichi-ban Edo kiriezu (mid-19th c.)

The continuity of history in the city of Tokyo is fundamentally different from that in Western cities. Tokyo poses a question to the world: What is history and spatial identity for each city?

In Tokyo, where the grand design relied on its topographic and natural conditions, the relationship between the road network and topography interests us first. In the high city, all extensive major roads are ridge roads while valley roads service local communities, with slopes branching between both types [fig.2].

The Imperial Palace sitting on the former Edo Castle site holds significance to the urban space of Tokyo. By retaining or transforming the topography to its advantage, Edo Castle, surrounded by the water in the moats and greenery, was nestled in an organically shaped site. The formation of the Imperial Palace as a legacy of Edo Castle can be seen to symbolize the organic formation of Tokyo as a whole.

Comparing an aerial photo of Ginza and Marunouchi with an Edo-period map, we can see that Tokyo has avoided such urban transformation experienced in Paris and Rome where historical continuity from an older urban structure was severed. Thus, Tokyo retains the basic structure of the city of Edo [fig.3].

In Tokyo, notably characterized by the lack of city walls, temple quarters tended to emerge along the city's fringes, aptly using the hilly landforms. As the city expanded, these temple areas formed semi-concentric rings like *Baumkuchen* [fig.4].

fig.6 大名屋敷を受け継ぐ庭園群（オーストラリア大使館、三井倶楽部、イタリア大使館）／Gardens with roots in feudal lord residences (Mitsui Club, the Australian Embassy, and the Italian Embassy)

いくのかがわかる。不動産関係の史料を比べてみよう。江戸の不動産史料の代表として、町人地の沽券絵図がある。一筆ごとの間口・奥行き寸法や坪数、売買価格、地主名、家守名などが記入されている史料で、建物には触れていないが、土地についてのソフト情報がある意味でイタリア以上に入っている。一方、建築に価値を置くイタリアでは、たとえば、ボローニャの修道院が所有する住宅群の16世紀後半の状態が詳細に図示されている。

それに対し、東京では建物は何度も建て替えられ、地上の視覚的要素は新しいものばかりだが、逆に、土地の仕組み（都市組織）は非常に古いのである。東京は江戸を下敷きにし、地面の上の世界を建て替え、更新しながら発展してきたといえる。

山の手を読む

山の手の都市空間の継続性をまず見てみよう。ヘリコプターから筆者が撮影した三田の元大名屋敷群である [fig.6]。現在、イタリア大使館、三井倶楽部が、地形を生かし、高台に建物を、南下りの斜面に池のある和風の回遊式庭園をもつ。もともとは湧水を生かした庭園で、今は、ポンプアップして水を供給している。東京は水資源が豊かな都市で、地図にも湧水が生んだ池がいくつもあり、そこから中規模な川が流れ出て江戸の都心に向かう様子が巧みに描かれている。

東京の山の手を調べていると、湧水の重要性が見えてくる。近代の都市開発で地下水のレベルが下がり、都心の湧水は減ったとはいえ、郊外、武蔵野、多摩にかけて今も東京に湧水は多い。それが神田川、善福寺川をはじめ中規模な河川となって都心を流れ、豊かなランドスケープを生み、都市のさまざまな機能、営みを支えてきた。

今回の対象地である谷中の「へび道」も、

Buildings and Land

A marked difference between Japanese and Western cities emerges in the perception of real estate. This is evident from comparing maps, for example, an 18th century map of Rome by Giambattista Nolli and a mid-19th century Edo district map (*Edo kiriezu*) [fig.5]; while Nolli's map highlights urban voids (plazas, streets, and courtyards) by differentiating buildings from open spaces, the Edo district map only illustrates lot and block boundaries (and no buildings). The Edo district map offers no indication of open spaces. Depicting little visual architectural data, the Edo district map offers much more land related data: land use (samurai area, merchant and artisan section, or temple and shrine district), family name of each residence, name of the merchant and artisan area, and temple names.

In contrast, in Italian cities where buildings constitute much of the real estate values, each building is numbered on a property map. The corresponding register lists the owner, area, and use of each floor. This shows how valuable buildings were and how this custom has continued.

Let us compare historical property records. A typical Edo-period property record is the picture-map of bills of sale (*koken ezu*) for a merchant and artisan area. The map contains no building illustration but much more land data than its Italian counterpart: each bill's width, depth, acreage, sales price, landowner, and caretaker. In contrast, Italy's property record depicts, for example, the housing conditions of an abbey in Bologna (*Abbazia dei Santi Naborre e Felice*).

In Tokyo, where buildings have been replaced many times, visual elements on ground are largely new, yet the urban fabric – the road network, district division, and lot configuration – is very old. In other words, Tokyo has developed by updating the world above ground, above the substratum of the city of Edo.

Reading the "High City"

Let us now look at the continuity of the urban space of the high city. The aerial photo I took from a helicopter captures a site in Mita where the residences of feudal lords (*daimyo*) stood [fig.6]. Today, the site houses the Italian Embassy and Mitsui Club. The establishments' buildings sit on a hill, ideally positioning walk-through landscape gardens typical of Edo on the south-facing slope. The lakes in these gardens originally held spring water but now hold pumped water. Tokyo boasts rich water resources, and this is evident from an illustration of the springwater ponds that generated mid-scale rivers flowing into central Edo.

Studying the high city reveals the importance of springs in Tokyo. Today,

fig.7 神田川沿いの目白斜面緑地。水の神のための聖域／
Sanctuary for a god of water, a green slope in Mejiro along the Kanda River
左／left
画像 ©2019 Digital Earth Technology, DigitalGlobe, The GeoInformation Group
地図データ ©2019 Google

fig.8 水神社と胸突坂／Water shrine (Sui-jinja) and a steep slope (Munatsukizaka)

水源は染井にある池から流れ出る藍染川の流路が1920年代に暗渠となったため、くねくねとへびのように曲がった形をしている。近代の画一化していく都市空間にあって、こうした暗渠のルートは、かつての都市の記憶をとどめる重要な要素だということが近年、認識されてきた。

　山の手の地形を生かした江戸の都市空間の在り方をよく受け継ぐ例として、目白台とその周辺の神田川沿いにある斜面緑地がある。南下りの条件の良い土地に、大名屋敷がいくつも並んでつくられていた。その名残として、緑に包まれたこのエリアには、湧水による池をもつ回遊式庭園が受け継がれている。神田川の守り神が水神社に祀られており、2本のイチョウが、ご神木として高くそびえる[fig.7、8]。

　外濠の外側に位置する市谷地区の高台には、江戸時代、奥行き20間のモジュールに基づき計画的、規則的につくられた下級武士の住宅エリアが広がっていた。敷地の前後で分割された所もあるが、今もその都市組織はおおむね受け継がれ、良好な住環境を維持している[fig.9]。

　じつは、起伏に富み緑あふれる山の手のタウンスケープの構成原理を分析していると、イタリア都市を対象に生み出された建築の類型 "tipo edilizio"（building type）と建築群から成る都市組織 "tessuto urbano"（urban fabric）を分析する方法だけでは不十分であることに気づく。東京の都市の本質を深く理解するには、緑の分布、聖域（寺社の境内）、道路網、敷地内のサイトプラン、さまざまな空地などが重要な要素となる[fig.10]。日本的な "Topos（トポス）" や "Genius loci（ゲニウス・ロキ）" はそれらの要素と深く関係している。それらも含んで都市を解析する必要から私は「空間人類学（spatial anthropology）」というネーミングを考え出した。

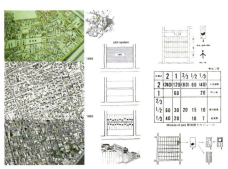

fig.9 市谷、下級武家地／Lower-class samurai houses in Ichigaya

fig.10 麻布の複雑な都市空間／Complex urban space of Azabu

central Tokyo has fewer springs than before as groundwater levels have lowered due to urban development in recent years. Yet, Tokyo has abundant springs on the outskirts from Musashino toward Tama. These springs originate the Kanda, Zenpukuji, and other mid-scale rivers streaming into central Tokyo, enhancing the rich urban landscape while supporting the city's functions and locals' activities.

Snake Street in Yanaka, this workshop's target area, meanders on the route of the Aisome River, which was channeled underground in the 1920s and flows from a pond in Somei. In recent years, amid more urban space being standardized, the importance of these channeled river routes have become more widely recognized as vestiges of old cityscapes.

Green slopes in the Mejirodai area along the Kanda River exemplify the preservation of the urban space of Edo where the topography of the high city was skillfully used. In this area, rows of feudal lord residences once stood on slopes with a desirable southern exposure. Today, these residences' walk-through landscape gardens fitted with springwater lakes remain as a legacy from the city of Edo. The area has a water shrine for a guardian deity of the Kanda River, with two sacred ginkgo trees standing high in the precincts [figs.7, 8].

During the Edo period, in the neighborhood of Ichigaya outside the Imperial Palace moat, a planned area of lower-class samurai houses, each measuring 20-ken (120-feet) deep, resided on a hill. While some lots have been subdivided into front and back

fig.11　不忍池と上野の山／Shinobazu Pond; Aerial view of Ueno's mountain
下／bottom
画像 ©2019 Digital Earth Technology, DigitalGlobe, The GeoInformation Group
地図データ ©2019 Google

fig.12　上野とその場所性。上野公園全図（1913）／Ueno Park map, depicting the characteristics of Ueno（1913）

2. 歴史的、文化的な資産を多く残すワークショップ対象エリア

　今回のワークショップでの対象エリアは、東京の北東部に位置しており、1980年代以後のグローバルな状況と連動した都市の華やかな開発、再開発がおおむね東京の南、西に集まるなかで、比較的安定した状態にあり、歴史的、文化的な資産を多く残し、日本らしさを生かした質の高い今後の都市開発、再生の可能性を大いに秘めているエリアだと考えられる。その点に注目し、「東京文化資源区構想（Tokyo Cultural Resource District）」という構想が近年提案されており、今回のわれわれが対象とするのもこの地域である。

上野エリア

　なかでも特異な在り方を見せる上野エリアは、不忍池と標高30mの上野の山が組み合わされ、世界の大都市でも珍しい大きな自然が都心に存在する[fig.11]。古代から日本人が好きな水の辺、山の辺を併せもち、山の上には古くから人が住み、古墳も多く存在し、中世から重要な寺院がつくられて聖地と見なされてきた。そこに江戸幕府の菩提寺である寛永寺がつくられ、江戸最大の聖域となった。その際に、京都が都市づくり（town planning）の手本とされ、江戸の北東、つまり鬼門の方向にある東叡山寛永寺は京都の比叡山延暦寺に、不忍池は琵琶湖に見立てられ、江戸に象徴空間がつくり出された。

　明治維新とともに天皇が江戸に移り、名称も東京に変わった。東京が近代国家・日本の首都となり、その新政府のもとで1873年、上野の寛永寺境内は近代の公園となった[fig.12]。山の上で国内の重要な博覧会が3度にわたって開催され、それを契機にこの空間に博物館、動物園、美術館な

portions, the area's zoning and lot division largely remain, offering a favorable living environment [fig.9] .

When studying the principles of the townscape formation of the hilly and green high city, we become aware that it requires more than just an analytical method, developed for Italian cities, that addresses the urban fabric (*tessuto urbano*) woven by clusters of buildings and building types (*tipo edilizio*) . The key factors to understanding the essence of Tokyo as a city include greenery distribution, sacred precincts (temple or shrine grounds) , the road network, site plans, and various open spaces [fig.10] . These factors are intertwined with the Japanese concepts of *topos* and *genius loci*. Recognizing the necessity to incorporate these factors into the study of the city of Tokyo, I coined the term "spatial anthropology."

2. Target Areas Characterized by Historical and Cultural Resources

This workshop's target areas located in the northeastern side of central Tokyo are stable compared with the southern and eastern sides of central Tokyo where development or redevelopment projects have flourished corresponding to the post-1980s global economy. Featuring historical and cultural resources, the target areas have potential for quality urban revival, taking advantage of their Japan-ness. Focusing on this potential,

the Tokyo Cultural Resource District vision, which covers the zone including the target areas, was proposed in recent years.

Neighborhood of Ueno

The neighborhood of Ueno is particularly unique among the target areas. With a 30-m high mountain and Shinobazu Pond, Ueno boasts a sizeable green spot rarely found in large cities across the world, offering both the waterside and hill districts favored by Japanese from ancient times [fig.11] . On the mountaintop, inhabited from long ago, old burial mounds exist. With important temples built from medieval times, the location has been seen as a sacred spot. In this location, Kaneiji – the temple of the Tokugawa Shogunate family in the Edo period – was built to become the then largest sacred spot. The planning of the Kaneiji temple modeled the town planning of Kyoto, likening the Kaneiji temple (on Mt. Toei) in the "demon's gate," or taboo quarters, to Kyoto's Enryakuji temple (on Mt. Hiei) and Shinobazu Pond to Lake Biwa. As such, the symbolic space was realized in the city of Edo.

As the Emperor moved to Edo during the Meiji Restoration, the city was renamed to Tokyo, the capital of the modern state Japan. In 1873, under Japan's new government, the Kaneiji temple precincts became a modern park [fig.12] . Japan's key exposition was held on the mountain top three times and a zoo, museums, and galleries were created to

fig.13 水の辺と博覧会（大正時代）／The waterside and expositions（1910s and 20s）

fig.14 上野の杜での花見／Cherry blossom viewing in Ueno Park

fig.15 アメ横／"Ame-Yoko," the Ameya-Yokocho market

どがつくられ、芸術と文化の公園としての機能をもつようになった。再び博覧会が明治末、大正の時期に上野で開催された際には、山の上に加え、不忍池の水の辺に第2会場が設けられ、夜のイルミネーションが水面に映えて大いに人気を集めた[fig.13]。上野の山の辺、水の辺は東京の近代文明を導入する窓口でもあった。上野の花見は今なお東京の人々にとって重要な祝祭のひとつとなっている[fig.14]。

　上野の山は文化の杜のイメージが強いのに対し、その下に降りると庶民の賑わいのある、小さな店が並んだ市場としてのアメ横（アメヤ横丁）がある[fig.15]。そのコントラストがいかにも東京的である。

本郷エリア

　再度、より一般的な大きな視点から東京の都市空間の構造を見てみよう。東京の前身、江戸は大きな城下町だった。身分制度に応じたゾーニングがあり、大名屋敷、中級武士の屋敷、下級武士の住宅、そして町人地として道路に面した表側に町家、路地裏に長屋という建築の明快なタイプが存在し、それらが集まって都市組織を形成していた。江戸城の東の平坦な土地には、規則的な都市組織ができる一方、南、西、北の起伏に富んだ山の手には、地形に応じた不規則な都市組織も広がった[fig.16]。

　今回のスタディ対象地のなかから、山の手の代表のひとつといえる本郷地区を例として見てみよう。丘の上の尾根を中山道（現在の本郷通り）が通り、それに面して加賀藩前田家の大名屋敷があったが、そこが今、東京大学のキャンパスとなっている[fig.17]。江戸時代の切絵図（古地図）を現在

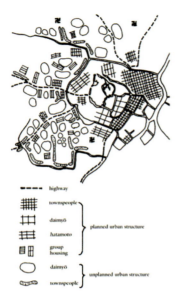

fig.16　都市組織のタイプ分類／
Urban fabric classification

fig.17　大名屋敷前田家と東京大学／
The University of Tokyo; The feudal lord residence of the Maeda family

shape the park of art and culture. When the exposition returned to the park at the end of the Meiji period and in the Taisho period, an area facing Shinobazu Pond was added to the original mountaintop venue. At night, illumination of the additional venue was mirrored on the water, and this gained much popularity [fig.13]. The mountain and waterside in Ueno served as a showcase introducing modern civilization in Tokyo. Today, enjoying cherry blossoms in Ueno continues to be an essential festive event [fig.14].

Unlike Ueno's mountain giving an impression of a cultural center, the mountain foot spreads the vibrant Ameya-Yokocho market that attracts commoners [fig.15]. This kind of contrast exemplifies Tokyo.

Neighborhood of Hongo

Let us look at the urban spatial structure of Tokyo once again from a broader perspective. Edo, the former body of Tokyo, was a large castle-town, where zoning corresponded to the class system. The urban fabric was comprised of clear architectural types: feudal lord residences, middle-class samurai mansions, and lower-class samurai houses; in the merchant and artisan area, tradesmen's houses lining streets and row houses along back alleys. A regular urban structure was formed on the plain east to Edo Castle while irregular urban structures responding to the hilly topography spread in the high city to the south, west, and north of the castle [fig.16].

fig.18 大名屋敷前田家とその周辺／The feudal lord residence of the Maeda family and its vicinity

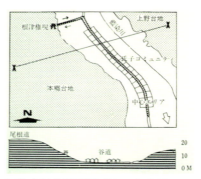

fig.19 台地と谷の都市断面／Urban section depicting plateaus and valley

の地図に重ねると、敷地の区画、道路網などが見事に一致する[fig.18]。交通上重要な広い道路は近代に拡幅されたが、それ以外の普通の道は、江戸のままといってもよい。東大キャンパスの構成は、前田家の大名屋敷の状態とは大きく変化しているが、赤門の場所、池の存在などは歴史を継承している。

　東京の都市の特徴を知るには、都市断面を考える必要がある。このあたりを断面で切ってみよう[fig.19]。本郷台地の前田家屋敷＝東大キャンパスを東に下りると、庶民的な雰囲気の根津の谷＝平地が広がり、その東の台地に登ると谷中の寺町につながる。根津では、町の領域の北側奥やや高い位置に、根津神社を千駄木から移動させた[fig.20]。祭りの日には、神輿が高い奥の神社から出て、下のコミュニティのエリアである町の中をパレードして、また上の神社に戻る。

　高低差は建築の類型にも反映され、大名屋敷が並んでいた本郷台地の上には、近代になって西洋風の建築が多くつくられ、一方、下の根津には、伝統的な庶民の家が存続し続けてきた[fig.21]。今も都市景観の違いが大きい。本郷台地から坂を西に下りた菊坂の周辺には、地形の変化に富んだ一角に、震災、戦災の被害を逃れた古い町並みが残る。井戸のある典型的な路地もあり、また階段状の坂に沿って木造3階建ての建物も残る[fig.22]。

下谷・根岸エリア

　上野台地の東下に広がる今回のスタディエリアのひとつ、下谷根岸については、1970年代の後半に陣内研究室で調査を行

fig.20　根津神社／Nezu Shrine

fig.21　伝統的な庶民の家（根津）／Traditional commoners' house in Nezu

fig.22　木造3階建の建物（菊坂）／Three-story wooden building in Kikuzaka

Take Hongo, a typical high-city neighborhood among the target areas. Along the Nakasendo (present-day Hongo-dori street) running upland stood the feudal lord residence of the Maeda family, where the University of Tokyo campus stands today [fig.17]. When this area's old Edo-period map (*kiriezu*) is superimposed on the current map, the lot division and road networks of now and then exactly correspond [fig.18]. The roads substantially remain from the Edo period, except important wide roads further expanded during modern times. The campus layout, while vastly different from the layout of the feudal-lord residence site, still retains the lake and the location of the Red Gate (*akamon*), inheriting the historical context.

To understand the features of the city of Tokyo, we need to consider its topographic sections. The section of the Hongo neighborhood shows that down to the east of the campus on the Hongo uplands spreads the Nezu valley, a plain brimming with commoners' qualities, and up to the east of the valley lies the plateau that houses temple quarters in Yanaka [fig.19]. In Nezu, Nezu Shrine was relocated from Sendagi up onto a slight platform deep toward northern Nezu [fig.20]. During a festival, locals carry a portable shrine, parading from Nezu Shrine down to and within the commoners' community in the town and then back up to the shrine.

fig.23 江戸の街道沿い町割りをとどめる明治時代の地籍図／Meiji-era cadaster depicting the zoning along the streets of the city of Edo

fig.24 小野照崎神社の富士塚。山開きの日／Mini-scale Mt. Fuji (*fujizuka*) of Onoterusaki Shrine, on the first day of the mountain climbing season

った。下谷・根岸は、日光裏街道に面して帯状に計画的につくられた町人地の構造が今まで受け継がれてきたエリアだ [fig.23]。日本の都市を構成する建築タイプが、都市のコンテクストのなかでどこにどのように立地し、近代化の過程でどんな構成の変化を見せてきたか、つぶさに観察できる。街道沿いには、商人の町家が並び、その間からなかに伸びる路地に面し、長屋が置かれる。こうした日常の空間が街道沿いに発達したのに対し、非日常の場である寺社はこの町人地の背後に潜んでいる。ここには、富士塚をもつ小野照崎神社が裏手にある [fig.24]。一方、上野の山に近い背後には、武家屋敷の系譜を引く庭付きの屋敷、独立住宅が並び、落ち着いた地区をかたちづくっていた。

町家、裏路地の長屋、武家屋敷の系譜を受けた独立住宅のそれぞれの建築類型を見ると、近代化の歩みをうまく吸収して発展したことがわかる。特に町家は看板建築化していくことで洋風のステータスを獲得した [fig.25]。多くの庶民が住んだ長屋がいちばんダイナミックな展開を示した。2階建てになり、水道とガスが引かれたことで、路地側にあった台所が背後にまわり、その部分が前室＝玄関の間に転じ、前面に格子を付けることも可能になった。内部を覗かれない構成も可能となり、長屋は近代に居住環境を大きく向上させた。

現在、背後の大きな屋敷はマンションに変わり、街道沿いの町家もだいぶ姿を消した。東京ではこうした歴史的な価値のあるゾーンを保存する制度が存在しないため、経済性を追求する都市の開発の流れのなかで、高層ビルへの建替えにより、古い建物が徐々に消えていく宿命にある [fig.26]。保存への努力も必要だが、同時に、都市の歴史的なコンテクストやコミュニティのつながりを大切にして、その土地にふさわしい建替え、地区の再生を行うことが重要な課題になっている。東京には今なお木造住宅

fig.25　街道沿いに並ぶ伝統的な町家／Traditional tradesmen's houses along the Nikko-urakaido route

fig.26　伝統的な町並みの景観を壊す高層マンション／Residential high-rise, disrupting the traditional townscape

The elevation gaps in these neighborhoods are reflected in architectural types. On the Hongo uplands, where feudal lord residences had stood, Western-style buildings were built in modern times; and in the Nezu valley, traditional houses of commoners continue to stand [fig.21]. Hence, the townscapes of Hongo and Nezu are dissimilar. In the Kikuzaka neighborhood down to the east of the Hongo uplands, an old townscape that escaped damage from the 1923 earthquake and 1945 bombings still remains in the corner on a versatile landform. The area has a typical alleyway with a well, and three-story wooden buildings continue to stand along a slope with stairs [fig.22].

Neighborhood of Shitaya-Negishi

In the late 1970s, Professor Jinnai's laboratory studied the Shitaya-Negishi neighborhood, this workshop's target area spreading on the eastern foot of the Ueno plateau. The Shitaya-Negishi neighborhood retains the structure of the merchant and artisan district planned along the Nikko-urakaido route in the Edo period [fig.23]. This neighborhood enables us to closely observe the location of the architectural types shaping Japanese cities and the structural changes of these architectures amid modernization. Tradesmen's houses lined up along the route, and row houses sat along its side streets. While this secular space developed along the route, a sacred space of temples and shrines resided to the back of the merchant and artisan district. One example is Onoterusaki Shrine that has mini-scale Mt. Fuji (*fujizuka*) in its grounds [fig.24]. In addition, detached houses and houses with a garden with the roots in samurai residences formed a peaceful area further back toward the Ueno plateau.

fig.27　ヒューマンスケールの都市空間（谷中）／Human-scale urban space in Yanaka

が多く、火災・地震にも強い住宅地にする必要もある。連続性を断ち切る通常の大規模開発ではなく、こうした歴史性、社会性を濃密にもつ都市のコンテクストに見合った適正規模の都市再生の手法を探求することが求められる。

谷中エリア

今回のスタディエリアのひとつ、谷中は1980年代中頃から、その歴史と自然、生活文化、コミュニティがよく受け継がれた地区として、東京のなかでも人気を集めている。この地区の再評価、イメージアップに貢献したのが、森まゆみら女性3人で始めた地域雑誌『谷根千』（谷根千工房、1984−）である。勉強会、街歩き、ワークショップを重ね、調査を深め、地域の歴史を掘り起こしながら、30年間、さまざまな特集を組んで発行してきた。

谷中には、特別重要な文化財、モニュメントはないが、地形の起伏に富み、地下水が豊富で緑も多く、寺院、墓地、坂、路地、井戸、長屋、屋敷など、多彩な要素の組み合わせによって、ヒューマンスケールの都市空間をつくっている。人々の生活がそこに感じられる [fig.27]。ドイツ人の日本学者、今井ハイデ（法政大学客員准教授）は谷中の路地に魅せられ、ノスタルジーではなく、路地の良さを今後に受け継ぐ方法を論じた本『東京の路地――移り行く都市空間の多様性と自在性』（Taylor & Francis、2017）を出版した。彼女によれば、谷中の路地の多くは、明治以後の近代化のなかで、それまで寺院が所有する大きな土地が細分化され庶民の住む宅地となるなかで、アプローチとして引かれたという。槇文彦、北山恒が論ずる「細粒都市」としての東京が近代に生まれたメカニズムがここにある。寺の土地の細分化は谷中の特徴だが、山の手全般では、大名屋敷が内部に路地を引き込み、多くの宅地のために細分化される傾向が随所に見られたのである。

These residential categories – tradesmen's, row, or detached houses – show their successful development, adapting to modernization. Specifically, tradesmen's houses developed into the early Showa-period billboard architecture, acquiring a status similar to that of Western architecture [fig.25]. The most drastic development was seen in row houses inhabited by many commoners, where a second story was added. With city water and gas lines laid, the kitchen facing an alleyway was relocated toward the back of the house and the former kitchen became an alcove to receive guests. This allowed a lattice to be fitted on the façade, an arrangement ensuring the occupants' privacy from the alleyway. As a result, the living environment of row houses greatly improved.

Today, multi-family buildings have replaced the large detached houses of the samurai residence lineage near the Ueno plateau and many of the tradesmen's houses along the Nikko-urakaido route are gone. With no scheme to conserve historic zones like Shitaya-Negishi, old buildings in Tokyo are destined to be replaced with high-rises amid an urban development trend that pursues economy [fig.26]. Just as efforts for historic zone conservation are vital, so is suitable urban redevelopment and revitalization that respect the historical context and local community. With clusters of wooden houses remaining, Tokyo is required to build

earthquake- and fire-proof residential areas. Considering these aspects, we need to search for a method of urban redevelopment at a scale suitable for the neighborhood with rich historical and social contexts, as opposed to the current large-scale redevelopment that severs the historical continuity of the local context.

Neighborhood of Yanaka

Since the mid-1980s, Yanaka, another target area, has become popular in Tokyo as a town with history, nature, lifestyle culture, and a good community. The local magazine *Yanaka, Nezu, and Sendagi* (commonly known as *Yanesen*), began by Mayumi Mori and two other women, contributed to the re-evaluation and improved image of the Yanaka neighborhood. The magazine was circulated for three decades with feature articles based on workshops and fieldwork to dig up historical stories in the neighborhood.

Despite having no important cultural properties or monuments, Yanaka is blessed with hilly topography and groundwater that keeps the area green. Its human-scale townscape consisting of various factors – temples, tombs, slopes, alleyways, wells, row houses, and residences – is brimming with a sense of locals' life [fig.27]. Having been captivated by alleyways in Yanaka, Heide Imai, a German Japanologist, published a book on a method to preserve good qualities

fig.28 谷中の典型的な長屋と路地（昭和初期）／Typical row houses and alleyways in Yanaka（early-Showa period）

fig.29 スカイザバスハウス／SCAI the Bathhouse

　震災、戦災で谷中は被害を受けずにすんだため、こうした路地に面する木造の長屋、仕舞屋が幸いたくさん残っている。だが、こうした伝統的なゾーンでも徐々に建替えが進んでおり、それにどう対応するかが問題となっている。

　長屋はすでに述べた通り、昭和初期まで空間の構成に発展が見られ、居住性がアップした。その優れた発展形の長屋の例を見ると、玄関には破風が付き、前室を簡単な塀で囲み、小さな庭をとっている [fig.28]。

　谷中のような庶民的界隈では多くの住民は銭湯に通うのが普通だったが、近代化でほとんどの家に風呂が付いたことで、銭湯の経営が成り立たなくなり閉鎖するケースが増えている。廃業した銭湯を現代アートギャラリーにコンバージョンした「スカイザバスハウス（SCAI THE BATHHOUSE）」は谷中にふさわしい、興味深い建築の活用例である [fig.29]。

　近年の動きとして、路地が多く歴史の香りをもつ谷中の雰囲気の良さを生かし、木造の住宅をリノベーションして洒落たショップ、レストラン、ワインバーなどが数多く登場している。いずれも小規模なものであり、谷中のヒューマンスケールを壊すことはない [fig.30]。そうした傾向は、前述のくねくねした面白い雰囲気のへび道界隈でも見られる。そこから分岐する路地にも、洒落た小さなバーが出現して若者を惹き付けている。谷中にはまた、町の魅力に惹かれ、クラフトの活動を行う若い人たちが外から流入し、工房や店を構える動きも見せている。

　近年の谷中には、もうひとつの注目すべき動きが生まれている。イタリアでは空き家の多い地方の歴史都市再生の切り札として、「アルベルゴ・ディフーゾ（albergo diffuso）」（散らばっている宿）と呼ばれる分散型ホテルの考え方が広がっている。受付、宿泊、朝食、夜の食事・交流の場などを町のなかに分散させ、町全体でホスピタ

fig.30 リノベーションされ店舗になった木造住宅／Wooden house renovated into a shop

of alleyways (*Tokyo Roji: The Diversity and Versatility of Alleys in a City in Transition*; 2017). Imai accounts that many of the alleyways in Yanaka were developed as approaches to commoners' housing lots that subdivided large temple sites. This elucidates the mechanism that made Tokyo into a "granular city" – a concept advocated by Fumihiko Maki and Koh Kitayama – in modern times. While the subdivision of temple sites was characteristic of Yanaka, the subdivision of feudal lord residences by letting through alleyways was typical in other places of Tokyo.

Having avoided damage from the 1923 earthquake and 1945 bombings, Yanaka still has wooden row houses and houses that used to have commercial spaces (*shimotaya*) along alleyways. Yet, the traditional zones, such as Yanaka, have been gradually exposed to redevelopment. The question here is how to address this change properly.

As mentioned before, the layout and living environment of row houses was upgraded until the early-Showa period. The better row houses have a gable above the main gate and a small yard surrounded by simple fences in front of the alcove adjoining the foyer [fig.28].

In the commoners' neighborhood such as Yanaka, locals had frequented a public bath, but as most houses became equipped with a bath amid modernization, more public baths are going out of business. Against this backdrop, SCAI the Bathhouse is an interesting conversion of a public bath into a contemporary art gallery, befitting the neighborhood [fig.29].

リティを発揮するというものである。それによく似たhanareという名前の分散型ホテルを建築家、宮崎晃吉が谷中に実現し、話題になっている。ここでは谷中の既存の町、コミュニティとのつながりを大いに生かす。風呂もは銭湯に行く。「町全体があなたのホテル」というコンセプトなのである。谷中の町も現在、あまりに観光客が増え、もともとの良さが薄れることを心配する声が強まっている。そのなかにあって、hanareは、谷中の中心からややはずれたあたりにあり、人々の日常性がよく感じられるエリアで、宿泊客にとっては最高の生活体験を得ることができる。同時に、歴史をもったコミュニティならではの高齢化、空き家などの問題を解決する優れた方法でもある。

　こうして既存のurban fabric 空間のスケール、人々の営みをうまく継承しつつ、現代に合わせた機能、活動を適切に導入することで、谷中の町は魅力をより一層高めているといえる。

むすび

　東京の北東部に位置する今回のワークショップの対象エリアは、幸いこれまで行政、民間ディベロッパーによる大掛かりな再開発をあまり受けてこなかった。そのため、歴史の蓄積を感じさせ、生活感もある個性をもった地域を育んでこられたのである。しかし、都心部に位置するだけに開発のプレッシャーを今後より強く受けるようになることが予想される。高齢化や防災の問題をはじめ、さまざまな今日的問題を抱えているに違いない。これらの場所にふさわしい、東京ならではの魅力ある都市空間をどう継承・創造できるのか。その可能性を3国の学生グループ、教員の間で大いに議論することは大きな価値を持つ。

fig.1　法政大学江戸東京研究センター蔵／Courtesy of Hosei University Research Center for Edo-Tokyo Studies
fig.2　陣内研究室作成／Source: Hosei University Jinnai Laboratory
fig.3　上空から見た東京中心部（筆者撮影）（上）、尾張屋版切絵図（19世紀中頃）（下）／（top to bottom) Aerial view of central Tokyo (courtesy of Hidenobu Jinnai) / Owariya Seishichi-ban Edo kiriezu [Sectional maps of Edo Compiled by Owariya Seishichi], (mid-19th c.)
fig.4　引用出典＝『Process Architecture／ETHNIC TOKYO』（プロセスアーキテクチュア、1987）／Process Architecture / ETHNIC TOKYO, Process Architecture, 1987.
fig.5　引用出典＝ M.Bevilacqua and M.Fagiolo eds., Piante di Roma dal Rinascimento ai Catasti, Artemide, 2012.／Source: M.Bevilacqua and M. Fagiolo eds., Piante di Roma dal Rinascimento ai Catasti, Artemide, 2012.
fig.6, 8, 10（下2点）、11（上）、15、17（上）、20-22、24-30　筆者撮影／figs. 6, 8, 10 (bottom)、11 (top)、15、17 (top)、20-22, 24-30 (courtesy of the author)
fig.7　引用出典＝ Google マップ（左）、参謀本部測量局5000分の1東京図（右）／ Source (left to right) : Google Map / Gosenbun-no-ichi Tokyo-zu sokuryo-genzu [The ordnance survey 1:5,000 map of Tokyo]
fig.9　引用出典＝『東京のまちを読む――都市・建築の構成原理に関する史的考察』（法政大学工学部建築学科建築計画研究室東京のまち研究会、1980）／ Tokyo no machi o yomu: Toshi kenchiku no kosei genri ni kansuru shiteki kosatsu [Reading the city of Tokyo: Historical study of the principles of urban and architectural formation], Hosei University, 1980.
fig.10　引用出典＝尾張屋版切絵図（19世紀中頃）（上）／ Source (top) : Owariya Seishichi-ban Edo kiriezu
fig.11　引用出典＝ Google マップ（下）／ Source (bottom) : Google Map
fig.12　引用出典＝『台東区史・社会文化編』（東京都台東区、1966）／ Source: Taitoku-shi shakai-bunka-hen [Social and cultural history of Taito-ku], Taito Ward Office, 1966.
fig.13　引用出典＝『Process Architecture／ETHNIC TOKYO』（プロセスアーキテクチュア、1987）／ Process Architecture / ETHNIC TOKYO, Process Architecture, 1987.
fig.14　撮影＝廣田治雄／Courtesy of Haruo Hirota
fig.16　引用出典＝拙著『東京の空間人類学』（筑摩書房、1985）／Source: Hidenobu Jinnai, Tokyo: A Spatial Anthropology, Chikumashobo, 1985.
fig.17　引用出典＝大熊喜邦『泥絵と大名屋敷』（大塚巧芸社、1939）（下）／Source (bottom) : Yoshikuni Okuma, Doroe to daimyo-yashiki [Distemper drawings and feudal lord residences], Otsukakogeisha, 1939.
fig.18　引用出典＝尾張屋版切絵図（19世紀中頃）／Source: Owariya Seishichi-ban Edo kiriezu
fig.19　陣内研究室作成／Source: Hosei University Jinnai Laboratory
fig.23　引用出典＝「東京郵便電信局地図」（1886）／Source: Tokyo yubin denshin-kyoku chizu [Map by the Tokyo postal telegraph office], 1886.

In recent years, wooden houses have been renovated to fancy shops, restaurants, and wine bars, taking advantage of the atmosphere of Yanaka being blessed with a historical townscape and alleyways. These new spaces are small enough to maintain the human-scale townscape[fig.30]. This trend in redevelopment can also be seen in the curious neighborhood of the aforementioned Snake Street. Along its side-alleys, small chic bars have popped up, thus attracting young people. Yanaka's appeal as a town has also invited flow of young artisans from outside who open their ateliers or shops.

Another recent move in Yanaka resembles a concept of "dispersed" hotels (*albergo diffuso*) spread in Italy as a means to revitalizing rural, historic towns that have abandoned houses. Dispersed hotels have spaces for reception, accommodation, breakfast, or dining and interaction spread in a town, offering town-wide hospitality to guests. *Hanare*, a dispersed hotel in Yanaka created by architect Mitsuyoshi Miyazaki, has attracted much attention. The hanare hotel brings out the strengths of the Yanaka neighborhood and the hotel's connections with it. For example, the hotel's bath is a public bath. The hotel's concept is "the town as a whole serves as your hotel." Today, as the number of incoming tourists has become excessive, locals of Yanaka have been expressing more concern over losing the qualities of their community. Yet,

the hotel near central Yanaka is located in an area brimming with locals' sense of daily life, offering guests immersion into an excellent local life experience. The dispersed hotel provides a good solution for issues of abandoned houses and aging population typical to historic neighborhoods.

These cases of redevelopment show that Yanaka is enhancing its appeal by aptly adopting contemporary functions and activities while maintaining locals' life and the scale of the urban fabric.

Conclusion

Fortunately, this workshop's target areas in northeastern central Tokyo have seldom been exposed to the large redevelopment led by government or commercial developers. Hence, the target areas have been able to nurture unique neighborhoods that offer senses of locals' life and history. Yet, amid central Tokyo, these areas are expected to confront more forces that push for redevelopment. In addition, these areas are faced with issues such as the community's aging and disaster prevention. How can we create, or preserve, appealing urban spaces only found in Tokyo that fit to these target areas? It is of great value to discuss possible solutions among student groups and faculty from the three countries.

上野公園を中心とした都市組織
The Urban Fabric of the Ueno Park Area

上野公園を中心とした周辺エリア／
6つの対象サイト

栗生はるか
（法政大学デザイン工学部建築学科教務助手）

不忍池・上野の森を抱く上野公園を、数百年にわたって存続する面的な広がりを持つヴォイドとして捉えてみる。この空虚な面に対して、呼応しながら、変容を繰り返してきた周辺一帯。それらをネットワーク化された都市組織として拾い上げることはできないか。ここは、高密な江戸の都市部にあって、庶民の生活に根差してきたエリアである。その中心には、長く寛永寺が鎮座してきたが、明治維新で博覧会の会場となったことをきっかけに祝祭・催事の舞台に変貌する。その文化的な影響は周囲にも波及した。界隈には多くの芸術家、文化人たちが生活をし、活発に交流が行われたという。

6つのサイトは、事前準備の段階で法政大学のSTA（special teaching assistant）の学生たちがそれぞれの興味、関心をもとに持ち寄った。選定されたサイトは、どれも異なる背景、個性、変遷を持ち、半径約1km圏内にありながら、多様な要素として混在している。急速に更新される都市、東京のなかでこの対象エリアの持つ意義は大きい。連続した歴史や文化、営みがどのように息づき、どのような問題を孕んでいるか。その今を綿密に解析し、これからを考えることは、東京という都市のあり方に多くの示唆を与える取り組みである。そしてそれは、その独自性を世界へ発信するヒン

トともなるであろう。

各サイトには「暗渠」「あんがわ」「つぶつぶ」等、それぞれの実態を読み解くためのキーワードが添えられた。それらはこのエリアに限らず東京の随所に秘められたメカニズムを説明するワードでもある。上野公園というヴォイドとともに変容を遂げてきた周辺エリアは、同じく中心に皇居というヴォイドを抱く東京の縮図のようにも映る。

フィールドワークでは、丹念な観察、分析から、それぞれの土地に内包されていた東京特有の複雑な成り立ち、そしてその可能性が浮かび上がった。ここで6つのサイトを順に紹介する。まず、上野公園の北西エリアに当たる、藍染川の「暗渠」、へび道の周辺。根津・千駄木界隈の生活圏として知られている。湾曲した川の形状に沿い、街並みから残されたポケットのような空地が特徴的である。そこから北上すると、長きにわたる市民活動がその風情をつなぎとめてきた谷中界隈（調査対象地域は厳密に言えば谷中・西日暮里だが、わかりやすさを優先して谷中とした）。昨今、観光化が進みつつあるが、「崖線」がつくりだす独特な景観と生活文化は柔軟に維持されている。南東へ下り、昔ながらの街並みと暮らしが残る下谷エリアへ足を伸ばす。「あんがわ」に喩えられる高層ビルに囲まれた木

The Ueno Park Area:
Six Target Sites

Haruka Kuryu

(Research Associate, Faculty of Engineering and Design, Department of Architecture, Hosei University)

Take Ueno Park, encompassing Shinobazu Pond and greenery, as an expansive areal void lasting through hundreds of years. Responding to this void, its neighborhood has had repeated changes. Can we recognize the Ueno Park area as a connected urban fabric? Located inside the built-up urban area of the city of Edo, this area has been rooted in commoners' lives, with the Kaneiji temple lying at its heart for generations. After the Meiji Restoration, the area served as an exposition venue and this led to its transformation into a place for celebration and events. The cultural impact spread to its surrounds, and the Ueno Park neighborhood was populated by a host of artists and intellectuals, promoting vibrant exchange.

During the workshop preparation, six target sites were selected by Hosei University students (special teaching assistants), following their interests. Clustering within the one-kilometer radius, each site with its unique background, identity, and changes adds to the versatility of the whole area. In the rapidly updated Tokyo, these sites hold significance. The essence of the workshop was in analyzing how the history, cultures,

and livelihoods in these sites continue to be alive as well as the issues involved, thereby envisioning futures. This task offers hints for a future of the city of Tokyo. These hints will then help expressing Tokyo's originality to the world.

For each site, a key word for interpreting the town was applied, such as *"ankyo"* (a covered or buried river), *"ann-gawa"* (a hard-edged block), and *"tsubu-tsubu"* (grains). These key words also describe the urban mechanism hidden in Tokyo at large. The Ueno Park area that has transitioned with the void – the park – can be considered as an epitome of the city of Tokyo where the Imperial Palace can be seen as a central void.

Based on detailed observation and analysis, this fieldwork shed light on the complex structure particular to Tokyo found in each site as well as the site's potential. Let me introduce the six target sites. The first is the area around Snake Street, tracing the buried Aisome River (*"ankyo"*), to the northwest of Ueno Park. This area, known as the living sphere of the Nezu-Sendagi neighborhood, features pockets remaining

造密集地帯を歩き回ると、公・私の境界が曖昧となった生活感あふれる情景に出会う。そこから少し南下し、関東大震災後に整備された産業道路が取り巻く東上野へ。度重なる街区の再編が「すきま」をつくりだし、小さいながらも多様な外部空間が存在している。高速道路を横切り南下すると、闇市から発展し、今なお活気あふれるアメ横の一帯が広がる。さまざまな担い手に委ねられたオープンスペースが、「つぶつぶ」な表情を見せている。そして最後に、上野の森の西側、本郷台地の大学機能に挟まれた弥生エリアの一部へ。積み「かさね」られた土地利用の変遷から、住宅地が奇妙な形の痕跡として残る。

　調査は、重層した都市組織から、かつての痕跡や人々の振る舞いをあぶり出すかのような作業であった。時代を超えて継承された小さなパブリック空間や、日常生活があふれ出る路地や空地などの都市の余白が多くのサイトで見出された。それらの余白は、その土地の歴史やコミュニティを物語る、地域の個性を許容している。6つのサイトの共通解ともとれるこれらの発見は、東京の魅力を支える重要な要素と言えそうだ。

　サイトによっては、大小さまざまな開発により今も大きくその様相が変化しつつある。地域性を顧みない開発に対しては歯止めをかけるべく、歴史が蓄積された建物・景観を新たな視点で生かし、地域のアイデンティティとして再編集してゆく動きも増えてきている。また、上野公園と周縁地域の関係性を再構築していこうという積極的な試みも起こりつつある。だが、それらは同時に観光客の集中やジェントリフィケーションを引き起こしかねないことも事実だ。実際、住人と観光客の共存を、目下の課題としているエリアも多い。

　周縁部は中心に比べ代謝が早い。その様相も抱える課題もスピーディに変わり続けている。そのような一帯だからこそ、積みかさねられた痕跡を、異なる多様な視点を通して丁寧に読み解くことが求められるはずである。その意味において今回の試みは有意義な作業となったのではないか。何を評価し、何を問題とし、何を抽出するか。異なる背景を持つ3カ国の教員、学生が江戸東京の都市組織の読解に挑戦をした。

along the street's meander. Heading north from Snake Street appears the Yanaka and Nishi-Nippori neighborhood where the traditional atmosphere has been retained through locals' efforts. Amid further transformation into a tourist destination, the distinctive townscape and the way of life shaped by the cliff edge (*"gaisen"*) has been maintained. Down southeast from there turns up the Shitaya neighborhood, which recalls the townscape and life typical in the old days. Walking around the dense area of wooden houses surrounded by high-rises – a configuration likened to a hard-edged block (*"ann-gawa"*) – takes us to a town-scape exuding a sense of the ordinary life where public-private spatial boundaries look smudged. Slightly down south from there is Higashi-Ueno, circumscribed by the industrial roads built after the Great Kanto Earthquake. Gaps (*"sukima"*) left out of repeated rezoing comprise small, yet various outside spaces. South across an elevated expressway pops up the Ameya-Yokocho market zone, which sprang from a black market and is still bustling today. The public spaces used at the discretion of various land owners or shop associations constitute the zone's granular impression (*"tsubu-tsubu"*) .

The last is a residential section in the Yayoi district to the east of Ueno Park. Through layering (*"kasane"*) of land uses, the section between university facilities of the Hongo uplands preserves a trace of the past in a peculiar shape.

The fieldwork can be described as digging up local behaviors and vestiges from the past out of the layers of the urban fabric. The remaining spaces of a city – such as the small public spaces existing through the ages, and the alleys and vacant spaces brimming with a sense of daily life – were found in many of the target sites. These remnant spaces allow the expression of the local identity that narrates the community and history of the site. This finding, which can be regarded as the common factor among the six sites, constitutes a key component of the appeal of Tokyo.

Some of the target sites are being changed by large- or small-scale development. To prevent urban development from ignoring local identity, increasing efforts are made to bring a new perspective to the use of the historic, built- and natural-scape so as to knit it into the local identity. In addition,

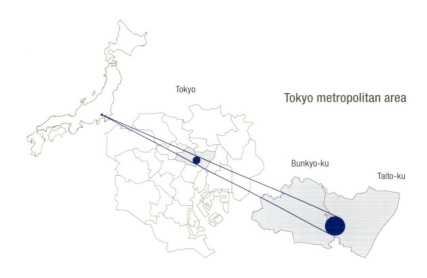

a proactive effort is gradually seen in re-organizing the relationship between Ueno Park and its vicinity. From a different perspective, however, these efforts may trigger gentrification or a flood of incoming tourists. In fact, some of the target sites are faced with ensuring harmonious living between locals and visitors.

Compared to the central area, its surroundings "metabolize" faster, and so do their appearances and problems. Therefore, studying the target sites naturally requires the close reading of the layers of traces on the land from multiple angles. In this point, this workshop should be meaningful. Challenged to identify the values, issues, and factors relevant for proposals from the target sites, the students and faculty from the three countries, with their unique backgrounds, strived to read the urban fabric of the city of Edo-Tokyo.

谷中
YANAKA

日暮里駅／Nippori Sta.

千駄木駅／Sendagi Sta.

谷中霊園／Yanaka Cemetery

下谷
SHITAYA

入谷駅／Iriya Sta.

根津・千駄木
NEZU & SENDAGI

鶯谷駅／Uguisudani Sta.

根津神社／
Nezu Shrine

根津駅／Nezu Sta.

弥生
YAYOI

上野恩賜公園／Ueno Park

上野駅／Ueno Sta.

東上野
HIGASHI UENO

東京大学／University of Tokyo

稲荷町駅／Inaricho Sta.

上野 アメ横
UENO AMEYOKO

本郷三丁目駅／
Hongo-sanchome Sta.

上野広小路駅／
Ueno-hirokoji Sta.

御徒町駅／Okachimachi Sta.

都市の時代変遷／Chronological Transition of the City

1859

出典：『5千分の1 江戸＝東京市街地図集成 1657年〜1895年』(柏書房、1988) ／ *Historical maps of Edo - Tokyo, 1657-1895*, Kashiwa Shobo, 1988

1880

出典：参謀本部測量局 2万分の1迅速測図／Source: Nimanbun-no-ichi Tokyo-zu sokuryo-genzu [The ordnance survey 1:20,000 map of Tokyo]

江戸時代から時代を超えて存続する要素から都市を読む
Reading the city from the elements retained from the Edo era

1932

出典：陸地測量部1万分の1東京近傍図／Source: Ichimanbun-no-ichi Tokyo kinbo-zu

1965

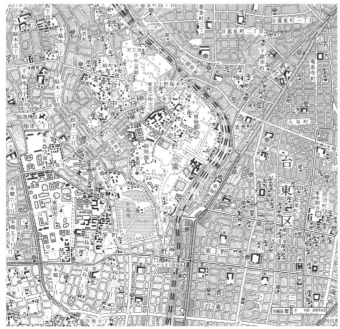

出典：国土地理院「地理院地図1万分の1」／Source: Geospatial Information Authority of Japan, 1:10,000 map of Tokyo

都市の構成要素／Urban Elements

Traffic
交通

The transportation network running in all directions

縦横に張り巡らされた交通網

- major road ／ 主幹道路
- subway ／ 地下鉄
- railway ／ 鉄道

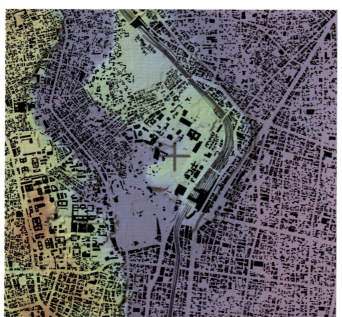

Topographical map
地形

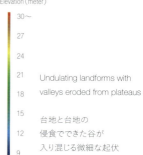

Elevation (meter)

Undulating landforms with valleys eroded from plateaus

台地と台地の
侵食でできた谷が
入り混じる微細な起伏

多様な要素から成る都市組織を分解して都市を読む
Reading the city by decomposing the complex urban fabric

Areal Void
面的ヴォイド

Voids existing for hundreds of years, such as public green spaces, parking areas, lakes, and temple and shrine grounds

公的な緑地や駐車場、
池や寺社地のような
数百年にわたって存続する
面的な広がりを持ったヴォイド

■ green space ／ 緑地　■ parking ／ 駐車場
■ cemetery ／ 墓地　■ water ／ 水面

Linear Void
線形ヴォイド

Voids, such as shopping streets and buried or covered rivers, underpinned by communities (which may include livelihoods) in the background

暗渠や商店街など
周辺エリアの生業やコミュニティを
背景に持つ線形ヴォイド

── buried river ／ 暗渠
── shopping street ／ 商店街

ANKYO

buried river

暗渠

「暗渠」とは、地下に埋設したり覆い被せたりして、上から見えないようにした水路を指す。東京を覆う水のネットワークを担っていた河川は、現在ではほとんどが暗渠化された。しかし、それでも「暗渠」となった道は、今も人々の生活を担う重要な軸となっていることが多い。「暗渠」は土地の記憶をひもとき、風景や地形を結び直す媒体である。

Ankyo refers to a waterway covered or buried underground, invisible from above the ground. Much of the river network in Tokyo has been converted into such waterways. The paths on top of these rivers often serve as axes essential for local life. Hence, these rivers are media with which to read the memories of the land and to reconnect the relationship among sceneries or landforms.

暗渠となった藍染川の一部「へび道」とその周辺／
Snake Street (tracing a part of the buried Aisome River) and its vicinity

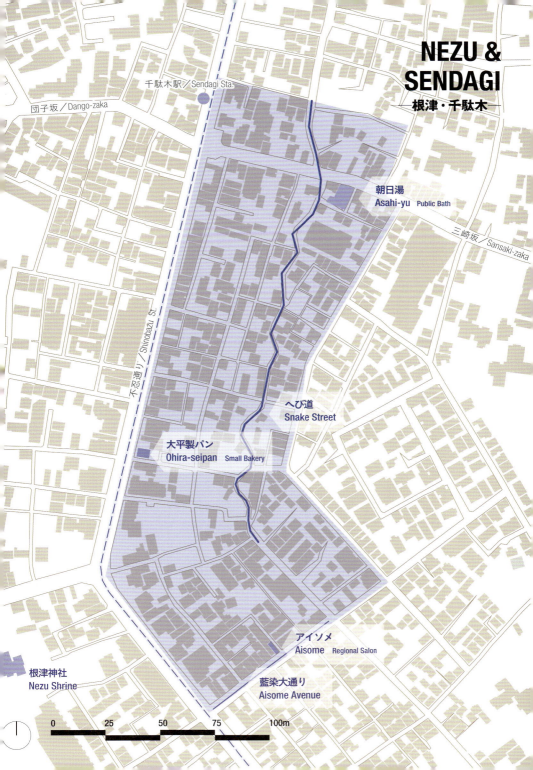

対象地は、今なお昔ながらの生活環境を残す根津、千駄木界隈と、観光化が進みつつある谷中界隈の境に位置する「へび道」である。かつての藍染川が暗渠化された場所に当たる。元来このエリアは湿地帯であり、藍染川は人々の生業を支えるための貴重な水資源とされていた。川は暗渠化され、13の湾曲を持つ「へび道」と呼ばれるようになる。道と建物の成立時代のズレから生まれる変わった形の空地、"変形空地"には、住民それぞれが街を住みこなそうとする工夫が表れている。また、接続する路地の周辺には、住宅と住宅に併設された多機能な場が点在している。かつての痕跡を生かしながら、生活者や来訪者にとってどのような場として機能するかが問われる一帯である。　　　　　　　　（安井勇吾）

The target site is Snake Street at the border between the Yanaka neighborhood, which is further developing into a tourist destination, and the Nezu-Sendagi neighborhood, which still retains the old-days' living environment. Snake Street is located above the buried Aisome River. The river area was originally wetlands, and the river was a precious water resource for livelihoods. After the meandering river was buried, the street above ground following thirteen curves became called Snake Street. The differences in periods between when the street was built and when buildings in its vicinity were built generated irregularly shaped gaps that reflect each local resident's ideas for enriching the neighborhood. Scattered around the street's alleys are multi-functional areas attached to houses. The question is what function the Snake Street neighborhood can offer locals and visitors. (Yugo Yasui)

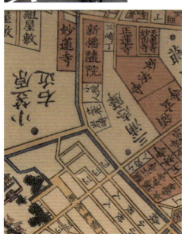

江戸時代の藍染川（1859年）／ The Edo-period Aisome River (1859)
出典：『5千分の1 江戸＝東京市街地図集成 1657年〜1895年』（柏書房、1988）／ Historical maps of Edo - Tokyo, 1657-1895, Kashiwa Shobo, 1988

「へび道」周辺の空地のプロット／
Part of Snake Street on the buried Aisome River

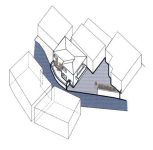  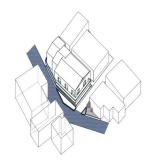

「へび道」と変形空地／Snake Street and irregular gaps

変形空地の使われ方／Use of irregular gaps

　調査にて、「へび道」の内部空間を詳細に観察し、ひもといてみた。第一に、藍染川によって形づくられた曲線沿いに、小さな空地、すきま、路地などの変形空地が残されていることがわかる。そこで、「へび道」と変形空地をつなげると、へび形というよりも、へびに"爪"をつけたような形が表れる。第二に、それら変形空地は一般的な道路沿いの空間とは異なり、より複雑な形状や機能、そして要素を内包していることが発見された。本調査では三次元モデルを用いて、「へび道」と変形空地とのつながり方について明らかにした。第三に、へびのようにうねっているため、道の湾曲した形状を、歩きながら視覚的かつ連続的に体感できることに気づく。このように、「へび道」の特徴はその形状だけではなく、連続する景観や残余空間と住人の日常生活との特殊なつながり方にあるということが明らかになった。　　　　　　　　（Chen Yue）

The student team closely studied the Snake Street space. Firstly, spaces shaped like pockets, gaps, and alleys remain along the meander. If these irregular spaces are connected to the street, the shape resembles not a simple "snake," but one with "claws." Secondly, the irregular spaces, unlike those spaces along ordinary roads, have more complex shapes, functions, and elements inside. Using a 3D model, the team interpreted how Snake Street links up with these irregular spaces. Lastly, the meander can be experienced visually and continuously while walking. Thus, Snake Street is characterized by its shape, continuous town-scape, and the relationship between the remnant spaces and local daily life.　　　（Chen Yue）

GAISEN   cliff edge

崖線

　崖地の連なりを示す「崖線」は東京の地形的な特徴のひとつである。大昔に台地が河川や東京湾の海の侵食作用により削られた痕跡である。自然の地形を残した「崖線」に広がる緑は東京の緑地帯の骨格となり、湧き水や動植物などの資源をたたえている。都市化が進んだ東京のなかで貴重な空間といえる。

Gaisen, or cliff edges, are a geographical feature of Tokyo left after plateaus were eroded by rivers or Tokyo Bay's ocean in ancient times. Retaining natural landforms, cliff edges (including their bottoms, terraces, and slopes) – enriched with natural springs, flora, fauna, and other resources – offer greenery that forms the foundations of green belts in Tokyo. In the urbanized city, these cliff edges are valuable spaces.

崖線と緑地の関係／
Relationship between the cliff edge and green land

谷中は上野の北西に位置する。対象エリアは谷中の地形的特徴が顕著な場所だ。東京という都市、特に山の手エリアでは武蔵野台地と窪地といった起伏が豊かな微地形により、多彩な環境が生まれている。神社や寺は緑の多い斜面のエッジに置かれ、聖なる空間として位置づけられた。そのような地形と建築の関係性は現在でも継承されている。斜面地においては坂や階段が数多く存在し、日常的に遠回りを強いられるなど、「崖線」に規定された独特の生活圏が広がっている。地図上では単調に見えるが、実際は地形に寄り添った非常に複雑な都市構造が存在している。

（近藤有希子）

The target site in the Yanaka and Nishi-Nippori neighborhood to the northwest of Ueno has prominent geographic features. In Tokyo, the "high-city" (*yamanote*) areas are particularly endowed with versatile environments attributable to undulated microtopography, such as the Musashino plateau and hollows. By the edges of green slopes, temples and shrines have been built, thus positioning the areas as sacred spaces. This landform-architecture relationship has been inherited even to the present day. Versants have multiple paths and stairs, forcing daily detours to locals. As such, unique living spheres unfold along cliff edges. Cliff edges, though monotonous features on a map, offer complex urban structures along their landforms. （Yukiko Kondo）

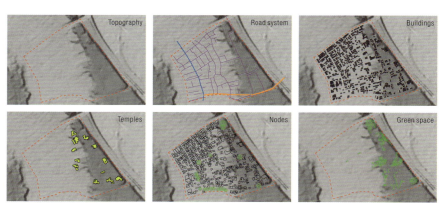

エリアの構成要素のプロット／Plot depicting the elements that comprise the area

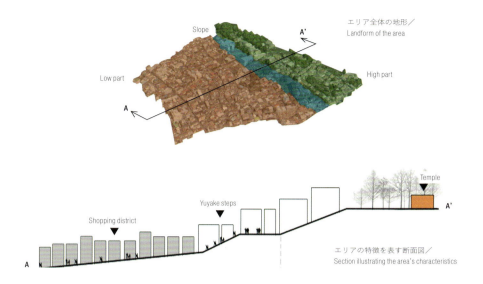

対象エリアは、高台に寺社地、低地に町人地という、江戸の典型的な空間構成のルールに従っているように見える一帯である。しかしながら、調査を通して見えてきたのは、地形に合わせた擁壁により、風景が段階的に変化し、独特な空間が生まれているという事実であった。谷中の地形に沿って歩くと、賑やかな商店街から静まり返った霊園へ、住宅街から寺社地へと変遷する。住民による鉢植えや花壇から人の手が加わっていない緑地帯へ、現実世界から精神世界へ。普段は見落としがちな変化が、この「崖線」に集約され、体験できる。また「崖線」が途切れた箇所には、生い茂った緑や空き家が見られ、時間が止まったかのように感じられる。それらは人為的な開発を防ぐ堤防のように存在し、都市のなかの貴重な居場所を守っているかのようだ。このような断絶を繰り返しながらもつながる谷中の存在はユニークである。東京において都市開発を議論するうえでの、重要な要素といえる。　　　　　（Nguyen Quang Tuan）

With a temple and shrine district on the hill and a merchant and artisan district on the valley, the target site appears to follow conventions of the Edo-era spatial planning. Yet, a survey done by the student team revealed that the scenery changes gradually from the lower to the higher area because of retaining walls tracing the landform, and this makes the Yanaka and Nishi-Nippori neighborhood a distinctive space. Walking along the landform takes us from an animated shopping street to a calm cemetery, to a residential area and then to the temple and shrine district. From potted plants and flower beds looked after by locals to a green belt of untouched nature; from the real world to the spiritual. This transition that tends to be overlooked in daily life is wrapped in the experience of this neighborhood. At dead ends along the cliff edge, rampant plants and vacant houses are found, appearing as if time has stopped, existing like a wall against urban development while preserving nature's rare habitat. Yanaka and Nishi-Nippori constitute a unique neighborhood where areas are connected through these transitions of time and space. This feature is a key element for discussing urban development in Tokyo.　　　（Nguyen Quang Tuan）

ANN-GAWA   hard-edged block
あんがわ

　江戸東京は、長らく大規模火災に悩まされてきた都市である。そのため現代では、消火活動や避難場所の確保のために、主要幹線道路の近辺は防火基準が厳しい。その代わりに、建ぺい率の緩和が認められ、高層の建物が多く建設されている。それらが耐火壁のようになり延焼を最小限に抑えると考えられている。つまり、東京では、通り沿いに高層ビルが立ち並び、その裏には小さなスケールの街並みが広がるという光景が多く存在している。その特徴的な様相を、まんじゅうの中身と外皮に見立てて、小さな街並みを「あんこ」、ビル群を「かわ」に喩え「あんがわ」構造と称することがある。

The city of Edo-Tokyo has suffered from large-scale fires for a long time. Thus, fire-prevention standards are higher near key arterial roads to facilitate firefighting and secure evacuation areas. In return, a lenient restriction on the building-to-land ratio is applied, leading to the erection of a host of tall buildings. These buildings are considered to stand as fire-proof walls, minimizing the spread of a fire. Thus, a townscape of high-rises along the road mixed with a small-scale neighborhood behind them is prevalent in Tokyo. This townscape structure is sometimes described as "*ann-gawa*," comparing the high-rises with the skin of a stuffed bun (*gawa*) and the small-scale neighborhood with its filling (*ann*).

木造密集住宅地（あん）とそれを取り囲む高層ビル群（かわ）／
Dense area of wooden houses (*ann*) and high-rises along its periphery (*gawa*)

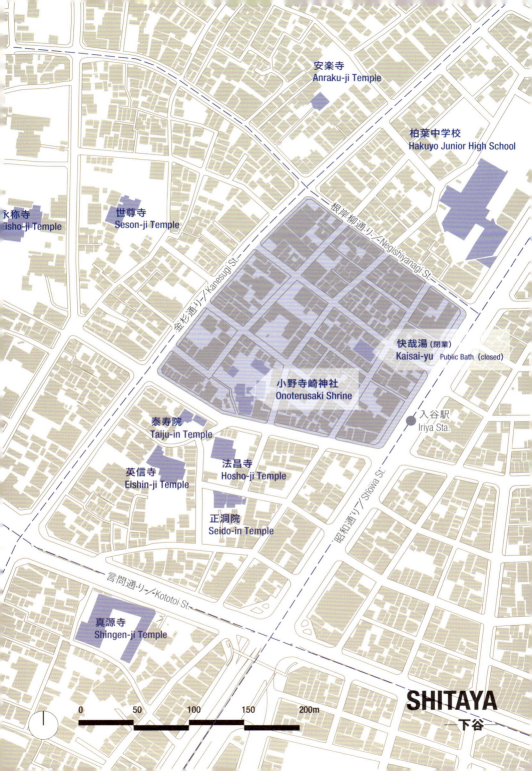

下谷の一帯は、旧甲州裏街道である金杉通り沿いが町人地として発達した地域である。その裏手には面的な広がりをもつ寺社が多く残る。関東大震災後には、復興道路として昭和通りがつくられた。主要道路がつくられると、通り沿いにあったかつての街並みの多くは、高層の不燃建築へと建て替わってしまった。だが、「かわ」である高層ビルの裏手には、「あん」に喩えらえる木造密集住宅地が残っている。そこには公・私の境目が溶けているかのような空間が生まれ、人々のおおらかな振る舞いが今も息づいている。建築物の密度感や、通りと住宅の関係性、コミュニティを支える祭りの存在などがこの地域の風情をつくり上げている。　　　　　　　（藤田彩加）

The target site in Shitaya sits in an area that had developed as a merchant and artisan district along Kanasugi Street (the former Koshu-ura-kaido byway). At the back of the site are many temples spreading their grounds. After the Great Kanto Earthquake, Showa Street, a major arterial, was created as a reconstruction road, and much of the then townscape along Kanasugi Street was replaced by nonflammable high-rises. Yet, behind these towers (*gawa*) remain densely-built wooden house areas (*ann*). In these areas, spaces with ambiguous public-private boundaries have been generated, where relaxed behaviors by locals are alive. These behaviors are fostered by the perception of building density, the relationship between the street and houses, and festivals that help knit communities.　　　　　（Ayaka Fujita）

高層ビル群は不燃の壁として存在する／
High-rises standing as nonflammable walls

高層ビル群の裏に残る木造密集住宅地／
Dense area of wooden houses amid high-rises

「ヘテロトピア」は世界に内在する"世界"である。外界と隔たりつつつながっている実在空間であり、外部の要素を内部に取り込んだり排除したりすることが可能な特殊な環境である。下谷の一帯は高いビル群の壁に囲まれ、東京という人口密集都市によって守られていると同時に、忘れ去られたかのような「ヘテロトピア」といえる。かつて木造密集住宅地は、建物と外部空間の境界が曖昧な都市組織を形成していた。下谷にはそのような当時の面影が残り、ほかの地域には見られない独特なヒューマン・スケールが息づいている。このため「あん」と外側の「かわ」の空間の対比が鮮明となる。本調査では、簡略的だが調整を施したビジュアルデータを用いてマクロ／ミクロのスケールを捉え、物理的な要素（都市・建築形態）と形を持たないソフト的な要素（社会・文化的側面）の関係を観察し、下谷に内在する特徴を分析した。

(Giorgia Greco)

Heterotopias are worlds within worlds. They are isolated yet connected spaces forming peculiar environments that can both include and exclude what is outside from what is inside[1]. Surrounded by walls of high-rise buildings, the target site in Shitaya seems both protected and forgotten by the overcrowded Tokyo: a heterotopia. Shitaya retains vestiges of densely-built wooden house areas from the past, where the urban fabric featured blurred boundaries between buildings and the outdoor space. Shitaya presents a peculiar human scale unlike in other areas. Hence, contrasts between the inner core and its periphery become evident.

Through a simple yet controlled visual narrative from the macro to the micro scale, the student team observed the relationship between the tangible sphere (urban and architectural forms) and the intangible sphere (social and cultural aspects) to analyze the intrinsic characteristics of the area. (Giorgia Greco)

★1　P. Johnson, Heterotopian Studies, [website], [n.d.], http://www.heterotopiastudies.com/ (Accessed 20 March, 2019).

SUKIMA   **gap**
すきま

「すきま」とは、わずかな間、空隙を指す。日本の都市には、かつての共同路地のほか、土地利用の変遷で期せずして生まれたもの、通風、採光、雨垂れや延焼防止などを意図して設けられたものなど多様な「すきま」がある。それらは、屋外へと延長される生活や生業を支え、抜け道や遊び場としても使われるなど、土地所有にかかわらず街の人々によって発見的に利用され、魅力的な外部空間を生む最小単位となっている。

Sukima refers to small gaps. In Japanese cities, various small gaps are found, such as: former communal alleys; unexpectedly generated spots as the land use transitioned; and spaces designed to let through the wind or light, or to prevent rain dripping or fire spreads. These gaps allow the extension of lives or livelihoods from the indoor space and can function as byroads or playgrounds. As such, with or without being owned, the gaps, incidentally used by locals, offer the smallest unit of space that provides an appealing outside space.

土地利用の変遷が「すきま」を生み出す／
Land use transition generates gaps

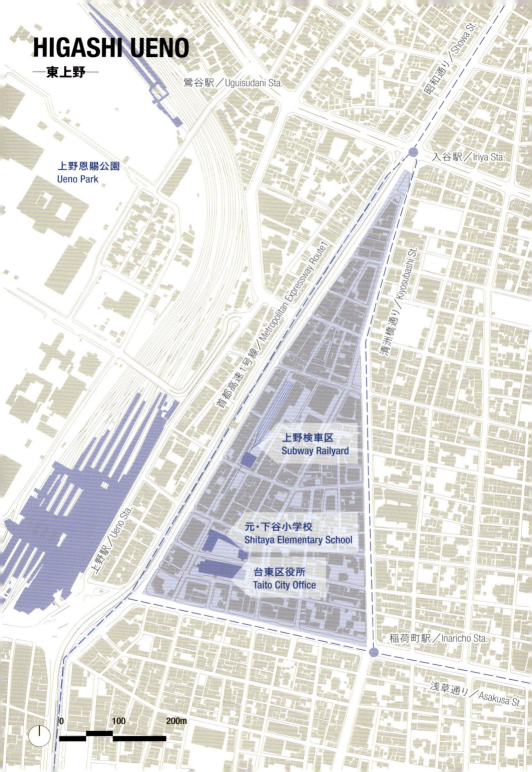

関東大震災後に整備された昭和通りを含む3本の産業道路に囲まれた東上野。江戸に起因する網目状の都市組織には高層の区役所や廃校となった復興小学校、東京初の地下鉄車庫が埋め込まれている。上野を支える各時代のインフラに小さな長屋などがいまだ隣接する雑多な風景からは、街全体の文脈を捉えにくい。だが、外部空間に着目して再び江戸時代からの都市組織の変遷を辿ると、道路開発や寺社の縮小によって、建物の更新に伴い街区の再編も行われたことがわかる。この流れの中で生み出されたヘタ地や、取り残された路地、開発で生まれた残地や空地など、現代の東京という都市が持つ小さな「すきま」にこそ、歴史の痕跡や人々の生きた振る舞いを垣間見ることができる。　　　（久保田啓斗）

Hemmed in by three industrial roads developed after the Great Kanto Earthquake, including Showa Street, Higashi-Ueno has a reticulate urban fabric, attributed to the city of Edo. In this urban fabric, a tall ward office, a now closed elementary school reconstructed after the earthquake, and the first subway shed in Tokyo are embedded. From Higashi-Ueno's mixed townscape where infrastructure from different periods abut small row houses, it is hard to understand the town's context. Yet, by following the past transition of the urban fabric from the Edo era with a focus on the outdoor space, it becomes clear that when the road was developed or when temple or shrine precincts were downsized, buildings were updated, and this entailed rezoning. Tokyo has small gaps – such as the irregular spaces or remaining alleys cut out by the rezoning, or the remnants or open areas left by urban development – which offer a glimpse into the vestiges of history or lively local behaviors.　　　(Keito Kubota)

豊かな外部空間のプロット／
Plot of versatile outdoor space

外部空間の使われ方／
Outdoor spatial use

　上野公園の東面に位置する当エリアは、一見、凡庸なる街並みに感じられる。本調査では、この一帯と公園を隔てて亀裂のごとく走る首都高との関係性を認識する必要があった。調査から、街の発展やインフラ整備の産物である対象エリアには、空間的連続性が存在しつつ、建物の高さ、材料、年代などがさまざまで、多様性のある都市構造が存在することが明らかになった。時代ごとに変化する土地利用が重層することで生まれた、細かな「すきま」や小道などのパブリックな空間は、(地―図の) 地の空間に存在するといえる。生活者たちは、それらの都市の残余空間に恣意的な機能を与えている。このように、対象エリアには異なる時代が混在し、個々の住人たちがバラバラに生み出した小さなパブリック空間がそこかしこに存在する。これらの小さな空間は多様であるが、一方である種同じ様相を呈しており、対象地のまとまった外観と印象を形成している。　（Monia Buongiorno）

Located to the east of Ueno Park, the target site appears to be a generic part of the city. For the purposes of the analysis by the student team, it was essential to detect the relationship between the site and the large highway to the west lying as a deep fracture between the site and Ueno Park. Through observation, the team found that the urban fabric of the site, shaped by the city's evolution and infrastructure development, is continuous yet non-homogeneous in terms of the building height, construction time, and materials. Shaped by the layering of land use patterns from different periods, public spaces, such as granular gaps or small paths, are found in negative spaces. These "micro public spaces" have functions arbitrarily endowed by locals. As such, the site looks like a mix of different ages interspersed with "micro public spaces" produced by local residents. These spaces are varied, yet in a way look homogenous and this homogeneity shapes the whole area's appearance and perception.

(Monia Buongiorno)

TSUBU-TSUBU **grains**
つぶつぶ

　たくさんの粒状のものが散らばるさまを「つぶつぶ」と称する。東京は、江戸からの変遷過程で土地が細分化され、建築物が密集した。そのため、それらは土地ごとに独自の発展を遂げ、その歴史やコンテクストを、多様な様相で表出させている。そのさまは、レム・コールハースに「ピクセルシティ」(『S,M,L,XL+──現代都市をめぐるエッセイ』より)、槇文彦に「細粒都市」(『TOKYO METABOLIZING』[TOTO出版、2010]より)と評された。

Tsubu-tsubu describes patterns made of dispersed grains. In Tokyo, land has been subdivided amid its transition from the Edo period, and buildings were densely erected. Hence, these lots developed uniquely, expressing their history or contexts in various modes. Referring to this phenomenon, Rem Koolhaas described Tokyo as a "pixel city" (*S, M, L, XL*), and Fumihiko Maki described it as a "granular city" (*Tokyo Metabolizing* by Kitayama, Tsukamoto, and Nishizawa).

土地ごとの独自の発展による多様な様相／
Various appearances resulted from unique lot development

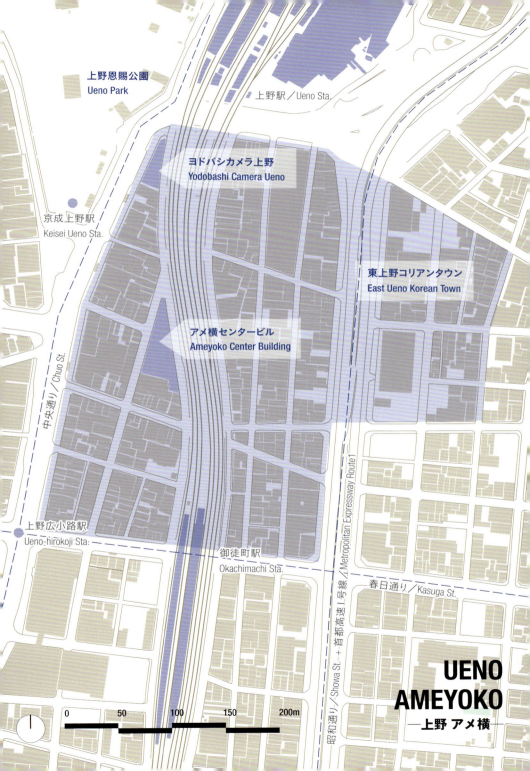

アメヤ横丁、通称「アメ横」は上野駅の南側に位置する。かつて、うねるように線路が通されたJR山手線により、奥まで見通すことができない空間が生まれた。その隠された空間に、戦後、違法にバラックが立ち並び闇市ができあがったという。現在も、多国籍な文化が入り混じる特殊なエリアである。そのエッジは交通量の多い道路や立体の高速道路により周辺と断絶されている。アメ横は、組織的なまとまりとして粒状に生まれた幾つかの商店会が集まることにより、大きな商店街が形成されている。各商店会は異なる"パブリック"を自由にデザインしているため、アメ横の景観は「つぶつぶ」な印象を持つ。

（磯目知里）

The Ameya-Yokocho market ("Ameyoko") sits to the south of Ueno Station. The curvy railroad of the JR Yamate line had left a pocket, where, after the war, illegal shacks stood in rows to form a black market. The black market district still constitutes an idiosyncratic multicultural spot, separated from its adjacencies by busy roads and an elevated expressway. The large "Ameyoko" shopping street is comprised of store associations that originally emerged as shops grouping as "grains" on a map. Each store association freely designs their public spaces, creating the granular impression of "Ameyoko."

(Chisato Isome)

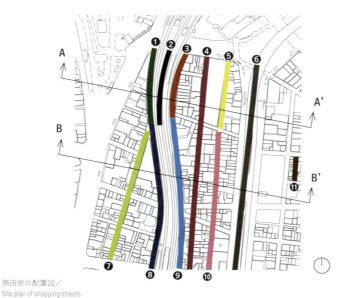

商店街の配置図／
Site plan of shopping streets

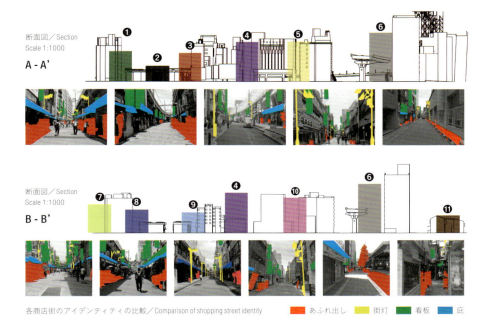

各商店街のアイデンティティの比較／Comparison of shopping street identity　　あふれ出し　街灯　看板　庇

対象エリアでは地形的な制約や歴史的な背景によって都市の骨格が段階的に形成され、商店街は駅を中心とし、道ごとに線形に発展したことが見てとれる。鉄道、高速道路、地下鉄といった公共交通機関はそのすきまを縫うように発展している。また、それぞれの商店街を観察すると、歩道の有無、街灯の装飾、商品のあふれ出し、看板の設置方法など、道ごとに異なる特徴があることが発見される。このことから、共有スペースの自由なデザインを許容する習慣が、非合法の商業空間である闇市時代から根付いていると考えることができる。つまり、公共空間としての道が各々の土地利用者、あるいは一連の団体によって個別のルールでデザインされているのである。道に対して生まれた民主的で自由なデザインは多様性を生み出し、「つぶつぶ」な都市空間をつくり出している。それらは、アメ横の魅力のひとつとなっている。　　（下平貴也）

The framework of the target site has been shaped in phases under the influence of geographical restrictions and the historical background. As we can see, the shopping streets grew from the station and along each street. The public transportation, such as railroads, an expressway, and subways, weaves its way north and south as if to run through gaps amid the shopping streets. From observation, particular characteristics of each shopping street are identifiable in the provision, or lack, of pedestrian path; street light ornament; goods "brimming" onto the street; and the method of signboard placement. This leads us to infer that leniency toward the free common-space use has been rooted since the black market period. To summarize, each street as a public space has been designed under the rules of each land user or association. Producing diverse appearances, the liberal, democratic design expressed onto the street shapes a granular urban space, which forms the identity and an attraction of "Ameyoko."　　(Takaya Shimodaira)

KASANE　　　　　　　　　　　　**layering**
かさね

　東京の都市空間は、過去から現在に至るまで繰り返されたさまざまな都市政策や土地所有の変化、地形造成などの事象が積層し成り立っている。江戸時代の切絵図を現在の地図に重ねると、敷地の区画、道路網など一致するものが多いことに驚く。「かさね」とは現在の東京をつくりあげてきた、社会的、地形的、空間的なレイヤーの重なりを示す江戸東京を読み解くひとつのキーワードといえる。

The urban space of Tokyo consists of layering of phenomena, such as land development and repeated changes in urban policy and land ownership. If Edo-period sectional maps are superimposed onto today's maps, surprisingly, much of the zoning and road networks coincide. *Kasane*, or layering, is a keyword for reading the city of Edo-Tokyo that signifies the social, geographical, and spatial layers that shape today's Tokyo.

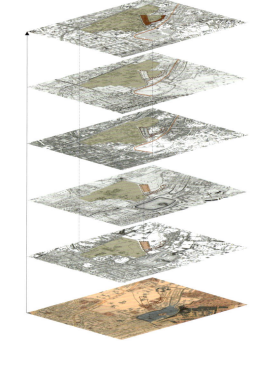

土地の変遷を表す地図の積層
出典:「5千分の1 江戸＝東京市街地図集成
1657年〜1895年」(柏書房、1988)、
参謀本部測量局2万分の1迅速測図、
陸地測量部1万分の1東京近傍図、
国土地理院1万分の1地理院地図
以上の図を加工して作成／
Layered maps highlighting land-use transition
Source: *Historical maps of Edo - Tokyo,
1657-1895*, Kashiwa Shobo, 1988.
Nimanbun-no-ichi Tokyo-zu sokuryo-genzu
[The ordnance survey 1:20,000 map of Tokyo],
Ichimanbun-no-ichi Tokyo kinbo-zu,
Geospatial Information Authority of Japan, 1:10,000 map of Tokyo

弥生エリアは不忍池の北西に位置し、本郷台地の地形的な起伏の境目にある。現在、台地には東京大学のキャンパスが広範囲に展開しているが、江戸時代には加賀藩、水戸藩などの大名屋敷地が連なっていた場所である。明治時代に入り、屋敷跡は新政府によって東京帝国大学（現東京大学）の用地として公収される。しかし、その敷地の一部は、地形的な特性から、警視局（現警視庁）の射撃練習場となり、明治10年に開場、後に皇宮地附属地の東京共同射的会社射的場となった。その後、射的場は移転したが、跡地は宅地化され土地の所有権は細分化された。そのため、射的場跡地は大学の敷地に挟まれ、不可思議な形で現在もその痕跡を残している。

（大澤岳史）

Located to the northwest of Shinobazu Pond, the Yayoi district borders on the Hongo uplands. On the uplands, where the University of Tokyo campus spreads, residential precincts of Kaga, Mito, and other feudal lords were established during the Edo era. In the Meiji era, the new government forfeited what used to be these feudal lord grounds to erect Tokyo Imperial University, today's University of Tokyo. Yet, due to geographical features, a part of the site was used for a firing range of the then Tokyo Metropolitan Police Department that opened in Meiji 10 (1877). Later, ownership of the firing range site was shifted to the Imperial Family and a shooting range of Tokyo Kyodo Shateki Gaisha was created. Subsequently, the shooting range was moved; the site was then turned into residential grounds and the ownership was subdivided. As a result, the former shooting range site, tucked between the campus grounds, still retains traces of its strange shape.

（Takeshi Osawa）

エリアの構成要素のプロット／Plot depicting the elements that comprise the area

住宅からあふれ出す植栽／
Potted plants "overflowing" from houses

コモンスペースになった隅切り／
Cut-off corners used as common spaces

通りに開いた社寺／
Temple or shrine precincts that open onto the street

住宅間のすきま／
Gaps between houses

　東京大学のキャンパスに挟まれたこの敷地が、開発を免れ、離れ小島のように住宅街として残ってきた背景には、所有権の細分化がある。所有権が細分化されたために、土地をまとめることが困難となり、開発の手が及ばなかったといえる。また、この一帯をくまなく歩き、さらに新旧の建物を微細に観察してみると、屋外空間の使われ方が、時代を超えて継承されている点に気付く。土地利用の変更の繰り返しのなかで生まれた余白に、植物があふれ出していたり、掲示板や消火栓・自転車などが置かれ、地域のちょっとしたコモンスペースになっていたりする。屋外空間の使われ方から、江戸から東京に変化する過程で起こった土地利用の変遷が、ゆるやかに連続し重なっている様子がうかがえる。このことからも本対象エリアは、さまざまな地形造成がレイヤー状に積み重なった、「かさね」によって特徴付けられているといえる。

（椿進之介）

Like a solitary islet, the residential area built on the former shooting range site escaped further development, and an underlying factor is the land ownership subdivision. This subdivision has hindered an acquisition of the whole residential area, presumably, preventing further development. By walking throughout the area and closely observing the old and new buildings, we notice that the outside spatial use has been passed on through the ages. The remnants attributable to land-use changes are sometimes serving as small common spaces, accommodating plants "overflowing" from the houses, noticeboards, fire extinguishers, or bicycles. This outside spatial use suggests the flexible layering of the land uses that have transitioned amid the city's shift from Edo to Tokyo. This provides support for the view that the target site features the layering, or kasane, of diverse land development.　　　　　　　　(Shinnosuke Tsubaki)

都市組織に挑戦する
Challenging the Urban Fabric

緑を探して
──東京の都市構造

クラウディア・カッサテッラ
(トリノ工科大学都市地域研究計画学部准教授［地域・景観計画］)

　東京では、緑の公共空間が一人当たり3㎡しかなく、ロンドン、パリ、ニューヨーク(各一人当たり26㎡、11㎡、18㎡)と比べて不足しています[1]。この都市組織のなかで、上野公園をはじめ明治時代につくられた大きな公園は島のように存在しています。島々の外側の地域はどうでしょうか。都内の緑の空間はどこにあるのでしょうか。丘や川、海辺など自然の風景を探して見つかるのは、コンクリートの擁壁、大小の水路、人工島です。

　観察する眼の焦点を一気に変えなくてはなりません。ヒューマン・スケールでは、道端の鉢植え、生い茂る植物、緑のカーテンなど、都市景観を彩る緑が見つかります。しかし、道路沿いに植栽があったり、通り沿いに並木があったりすることは稀で、並木通りであったとしても、植木を棒のようにしてしまうのはなぜかしらと首を傾げてしまうほどに剪定されています。

　そんななか目に飛び込んできたのが、神社の杜です。神聖な木、岩や湧き水もあります。寺社の杜は東京の都市構造のそこかしこにある身近な存在で、敷地内で休憩をとることができます(が、腰掛けられるベンチはなく、午後5時には閉まることもあるのでご注意ください)。やっと見つけた東京の自然は、鳥居の奥の神道の領域とい

うわけです。

　裕福な人は、住宅地の外壁の奥にある私有地の庭に緑を有しています。庭をもたない人は、家の前というややパブリックな場を緑の空間にしています。隅っこや狭いすきま、玄関前の鉢植えがその例です。子どもの遊び場はその大半を四角い砂地が占めており、緑の空間と呼ぶに値しません。

　オフィスや多様な用途の建物が混在する街区で構成された現代風の地区に入ると、いわば現代性の象徴や、ディベロッパーによる寛大な取り計らいの象徴としてすきまの空間が存在しています。こうした空間はビジネス仕様で、使うためではなく見てもらうためにあります。都市のすきまは避難所や延焼防止の緩衝帯として役立つ貴重な空間です。また、自然は必要とされず、スポーツ施設が歓迎されています。昨今の条例では、公園内での商業行為が推奨されている場合もありますが、この施策が公園の自然に人工的な要素が加わる素地となる可能性があります。

　これまで説明してきたタイプの緑が、都市および地区のスケールで東京の都市構造に一定のパターンを生み出しています。西欧人、殊にイタリア人である筆者は、緑の空間が不足していることに驚くと同時に、思いがけず目に入る植物、花、生い茂る植

Searching for Green Patterns within Tokyo's Urban Fabric

Claudia Cassatella

(Associate Professor of Regional and Landscape Planning, Interuniversity Department of Urban and Regional Studies and Planning [DIST] , Politecnico di Torino)

Tokyo is particularly lacking in public green spaces, when compared to cities such as London, Paris or New York: only 3 ㎡ per capita, against 26, 11 or 18 ㎡ , respectively[1]. Some big public parks, created during the Meiji Era, such as Ueno Park, stand like islands in the urban fabric. What about the rest of the city? Where are green areas in Tokyo? When looking for natural features, such as hills, rivers, and the seaside, somehow concrete retaining walls, channeled rivers and streams, and artificial islands are found instead.

A jump in the scale of observation is needed. At the human scale, the eye catches several green items in the urban landscape: vases on the sidewalks, rampant plants, green screens. Rarely, roadside greenery or a tree lined avenue (but branches are cut in such a way that one could wonder why planting a tree and then reducing it to a wood pole) .

Then, suddenly, a wood comes into the eye. It's a shrine, its garden and a sacred tree, rocks, and even spring water. And, there are sacred woods scattered in the urban fabric, just around the corner. And you can go in

and have some rest (be aware there are no benches, and it closes after 5 pm). So, finally, nature is there, in the Shinto territory, beyond the Torii gate.

Or, (for affluent people) it's in private gardens beyond the walls in residential neighborhoods. Or, for those who don't have a garden plot, nature is present in the semi-public space in front of their house. A corner, a narrow strip, or a vase right in front of the door. It's not worth counting in children's playgrounds as nature, because they are sandy squares for the most part.

Coming to the contemporary districts made by offices and multipurpose blocks, open spaces are offered as a sort of icon of modernity and generosity by the developer. Corporate style, something to be seen, not to be used. Open spaces are valuable as "evacuation areas" or fire disaster prevention buffer zones. Nature is not required, sporting fields appreciated. Recent legislation encourages business activities inside existing parks, a possible premise for further artificialization.

These are the types of greenery that create

生から受ける刺激に嬉しさを覚えます。おっと、土砂降りの雨です。雨宿りする場所を探さなくては。その後、ほどなくして陽が射し始め、暑くなってきました。今度は冷房の効いた屋内を探さなくては。そうこうしているうちに、ふと思うのです……。

　サステイナビリティの概念はどうなっているのだろう？と。林冠や公園の一人当たりの面積、種の多様性などの一般的な指数を見ると、都内にある緑の空間のパターンでは効果が薄そうです。都内の緑の空間パターンには地図に載らないタイプがあるため、緑の空間の規模を測ることができないのですが、ランドスケープ的連続性や多様性などの生態環境の観点から見た緑の空間の価値という意味では、疑問符が付きます。

　緑のインフラという概念を東京をはじめとする都市圏に当てはめると、一風変わった意味合いを帯びてきます。東京はもともと開かれた都市で、壁面・屋上緑化、雨水を利用した庭などのハイテクな解決策の開発や試験導入を行う態勢があります。しかし、浸水性のある土壌、小川やのびのびとした植生など、自然がつくる生態系には居場所がありません。自然は神社領か植木鉢のなかにある。これが東京で得た大いなる学びです。これはしかし、本当に遥か遠い日本に限られた現象でしょうか。

★1　Tokyo 23 wards（Data source: MILT & TMG, elaboration by A. Iida The University of Tokyo）（Accessed March 20, 2019）／「都市公園」および「都市公園以外の公園」の総面積は一人当たり5.70㎡である（東京都建設局HP「Parks in Tokyo」[http://www.kensetsu.metro.tokyo.jp/english/jigyo/park/01.html] 2019年3月20日最終確認）。

84

patterns into the city fabric at the district and urban scales. A Western eye, an Italian one, is at the same time staggered by the lack of green spaces and delighted by the sensorial stimuli provided by sudden revelations of plants, flowers, and lush vegetation. Then, it heavily rains, and she searches for a shelter. A moment later, it's sunny again, and hot, and she searches for a shelter again, and for air conditioning. And then she realizes.

But...what about environment sustainability? Tokyo green patterns are unlikely to be functional according to well-known indicators, such as tree canopy, park land per person, or number of species. Some of the green features cannot even be on the map, so they cannot be assessed at the urban scale, and their ecological value is questionable (in terms of landscape connectivity, heterogeneity, etc.).

The concept of green infrastructure, when applied to a city region such as Tokyo, takes a peculiar shade. Tokyo is naturally open, and ready to test and invent hi-tech solutions, such as green walls, green roofs, or rainwater gardens. Ecosystems made of natural processes (permeable soil, natural streams, spontaneous vegetation) cannot find a home. Nature dwells in a sacred precinct or in a vase. This is the great lesson learnt from Tokyo. But, is it really so far and so exotic?

★1　Tokyo 23 wards (Data source: MILT & TMG, elaboration by A. Iida, The University of Tokyo). The total area of urban parks and "parks other than urban parks" is 5.70 square meters per capita, according to Tokyo Metropolitan Government Bureau of Construction, Parks in Tokyo (http://www.kensetsu.metro.tokyo.jp/english/jigyo/park/01.html, accessed 20 March, 2019).

Photos courtesy of Claudia Cassatella／
以上すべて筆者撮影

東京の見えない空間

ニコラ・ルッシ
(トリノ工科大学建築デザイン学部准教授[建築意匠])

『東京の空間人類学』[★1]における陣内秀信の考察や、『ペット・アーキテクチャー・ガイドブック』[★2]や『メイド・イン・トーキョー』[★3]などアトリエ・ワンの2000年代以降の研究成果は、鈴木明曰く「伝統的な都市計画の観点から説明するのが難しい都市」[★4]である東京を読み解くうえで今もなお有効である。

レム・コールハースは、ロラン・バルトが『表徴の帝国』において著した都市の複雑さについて2011年に再び着目し、東京は「中心が空白の巨塊(ヴォイド)」[★5]であると述べた。都市を単純化したかたちで捉えることにより、都市類型の尺度の両端に位置する2類型を把握することが可能となった。しかし、絶えず変化し、本来複雑で捉えがたい都市構造をより詳細に観察するうえでは有効なアプローチではないことが明らかである。

東京の都市構造の発展の経緯については陣内秀信の著書に詳述されているのだが、古くからある都市分類法の基準では測れない論理、ヒエラルキー、構造や素材によって発展したのである。アトリエ・ワンの研究は設計の結果として生まれるすきまに着目し、高尚な建築("high" architecture)とその対極にある建築("low" architecture)の間にある伝統的な質的ヒエラルキーを拒否し、空間の利用に主眼を置いている。「ペット・アーキテクチャー」や「ヴォイド・メタボリズム」など日本の都市について語るときに用いられる専門用語から成る新たな都市学のコトバの定義の枠組みは、正確かつ綿密な観察によってつくられるのである。

新しい都市学のコトバによって、稠密と空虚、パブリックとプライベート、建築空間と公共空間という二分律を乗り越えることが可能となり、極端に異なった特徴をもつものの、特定の形、境界、機能を備えたタイプの空間を地図上に示すことが可能と

  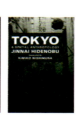  

Roland Barthes, *The Empire of Signs*, 1970. ／ロラン・バルト『表徴の帝国』(1970)
Rem Koolhaas and Hans-Ulrich Obrist, *Project Japan, Metabolism Talks…*, 2011. ／レム・コールハース+ハンス=ウルリッヒ・オブリスト『プロジェクト・ジャパン――メタボリズムは語る』(2011)
Hidenobu Jinnai, *Tokyo : A Spatial Anthropology*, 1995. ／陣内秀信『東京の空間人類学』(1995)
Hidenobu Jinnai, City diagram from *Tokyo: A Spatial Anthropology*, 1995. ／街のダイアグラム(陣内秀信『東京の空間人類学』[1995])
Tokyo Institute of Technology Tsukamoto Architectural Laboratory & Atelier Bow-Wow, *Pet Architecture Guide Book*, 2001. ／東京工業大学建築学科塚本研究室+アトリエ・ワン『ペット・アーキテクチャー・ガイドブック』(2001)

Invisible Spaces in Tokyo

Nicola Russi

(Associate Professor of Architectural Design, Department of Architecture and Design [DAD] , Politecnico di Torino)

The reading given by Hidenobu Jinnai in *Tokyo : A Spatial Anthropology*[1] and the subsequent research published since the 2000s by Atelier Bow-Wow, such as *Pet Architecture Guidebook*[2] and *Made in Tokyo*[3] are still an effective way of interpreting the city of Tokyo which, as Akira Suzuki says, is "difficult to describe in terms of traditional urban planning."[4]

An urban complexity, synthesized by Roland Barthes in *The Empire of Signs*, was later taken up in 2011 by Rem Koolhaas describing Tokyo as "a mass surrounding a central void."[5] This reduction of the city to a simplified figure has made it possible to recognize a clear dichotomy between the two opposites on a large scale but it has proved ineffective in a closer observation of an intrinsically complex, kaleidoscopic and elusive urban fabric. Tokyo's urban fabric is a particular texture, which is described in Jinnai's text through its historical evolution. This texture has developed with the logic, hierarchies, structures and materials that are beyond the classical criteria of urban taxonomy.

Atelier Bow-Wow's research, which focuses on the interstices created by planning, rejects the traditional qualitative hierarchies between "high" and "low" architecture and focuses primarily on the uses made of these places. Precise and minute observations form the framework for the definition of a new urban vocabulary which consists of a terminology specifically adopted to the Japanese city such as "pet architecture" or "void metabolism."

New descriptive methods that have allowed us to overcome the dichotomy between full and empty, between public and private, between built and public spaces have made possible a mapping of spaces with extremely different characteristics, but all characterized by having precise forms and boundaries and specific functions. Almost two decades after those first analyses of Tokyo, is it now possible to expand these surveys to recognize other forms of spatiality and further themes of research and investigation?

From the observation conducted during the workshop "Challenging the Urban Fabric," open spaces of extremely different nature have been identified, but they all are characterized by a certain degree of functional indeterminacy and by having boundaries that are difficult to circumscribe. Unlike "pet architecture," these open surfaces, almost invisible to an initial planimetric observation of the city, reveal

Ordinary townscapes of Tokyo, photos courtesy of Nicola Russi, 2018／東京に遍在する街の風景（筆者撮影、2018）

なった。東京についての初期の研究から20年ほどが経った今日、その研究結果を応用して新たな空間性の形態を見つけ出し、研究テーマを深めることができるだろうか。

「都市の文脈に挑戦する」と題した本ワークショップで行った実地調査から、性質が大きく異なるすきまの空間が特定された。しかし、これらのすきまは一様に明確な機能や境界をもっていない。ペット・アーキテクチャーと異なり、東京を平面図で見たときにはあまり見えてこなかったすきまだが、ワークショップの結果は東京の高度に構築された公共空間システムにおいてすら変わりやすい物理的・生態系的特徴とともに残る曖昧な空間利用があることを露呈している。

歩道の公共空間と建物が接する箇所では、概して幅30cm以内の線形のスペースが家庭と都市生活の緩衝地帯として働き、多様な活動の場となっており、狭い小庭、自転車置き場、雑多な物置場、もしくは公共または私用の設備（水栓や自販機）の敷設場所などとして利用されている。ペット・アーキテクチャーは「東京の大きさとヒューマン・スケールを取り持つ」[6]存在であると貝島桃代は述べているが、ペット・アーキテクチャーと同様にその線形空間のネットワークが、活気ある地区を形成し、居住性が高く、他者を呼び込む都市をつくるための貴重な役割を果たしており、主人の自発的かつ気ままな空間利用を容認している。

谷中の斜面にある空き地や放棄された建物には、形、大きさ、そして地形の3つの点で特質がある。コンパクトな都市構造に囲まれており、都市再生のプロセスから取り残されているこの珍しい場所に意外にも出現した都市の〈原野〉があり、自然が建物に侵襲して、建物は周辺と隔てる仕切り板もなく周囲の空き地に開放されている。

アトリエ・ワンによって定義づけられた建築概念とは異なり、これらのすきまは、形、プログラミング、空間利用の観点から見ると残余の空間である。すきまは空虚な線、点、面であり、東京という都市に活気を与えている。これらのすきまはその機能が未定で、可能性を秘めた貴重な空間であり、今日計画できない街の姿と都市の原動力を下支えする空間として存在している。

★1 Hidenobu Jinnai, Trans. Kimiko Nishimura, *Tokyo: A Spatial Anthropology*, University of California Press, Los Angeles, 1995.
★2 東京工業大学建築学科塚本研究室＋アトリエ・ワン『ペット・アーキテクチャー・ガイドブック』（ワールドフォトプレス、2001）
★3 貝島桃代＋黒田潤三＋塚本由晴『メイド・イン・トーキョー』（鹿島出版会、2001）
★4 Akira Suzuki, *Do Android Crows Fly Over the Skies of an Electronic Tokyo? The Interactive Urban Landscape of Japan*, Architectural Association, London, 2001, p.48.
★5 レム・コールハース＋ハンス＝ウルリッヒ・オブリスト『プロジェクト・ジャパン メタボリズムは語る…』（平凡社、2012）12頁
★6 Nicola Russi, *Backgrounds*, Quodlibet, Macerata, 2018, p.127.

how even in Tokyo's hyper-programmed system of public spaces, islands of ambiguous use survive with variable physical and morphological characteristics that are difficult to circumscribe.

In the contact points between the public space of the sidewalk and the buildings, linear surfaces that often do not exceed 30 cm in width host numerous activities acting as a buffer between domestic and city life: a small linear garden, a bike parking, a storage for various materials or private and public equipment such as water taps and vending machines. Similar to the "pet architecture" described by Momoyo Kaijima as objects of "mediation between the greatness of Tokyo and the human scale,"[6] the network of these thin surfaces that animate many neighborhoods of the ordinary city, fulfills the precious task of making the city more welcoming and livable allowing spontaneous and unplanned uses to the citizens who live there.

The system of open spaces and abandoned buildings that has been mapped on the slopes of the hill in the Yanaka district is of an extremely different nature, both in terms of size and form and for the exceptional topography that distinguishes them. In this uncommon place, which is enclosed within a compact urban fabric and separated from its processes of regeneration, an unexpected urban "wilderness" takes place where the nature penetrates the buildings and they open without diaphragms to the open space that surrounds them.

Unlike those identified by Atelier Bow-Wow, these interstitial spaces are residual in form, programming and use. They are lines, points and surfaces of emptiness that animate the city of Tokyo, endowing it with an almost invisible system of precious potential places of indeterminateness, to be preserved as a support for practices and urban dynamics that are impossible to determine or plan in our present times.

[1] Hidenobu, Trans. Kimiko Nishimura, *Tokyo: A Spatial Anthropology*, University of California Press, Los Angeles, 1995.
[2] Tokyo Institute of Technology Tsuamoto Arch. Lab & Atelier Bow-wow, *Pet Architecture Guide Book, Living Spheres Vol. 2*, World Photo Press, 2002.
[3] Momoyo Kaijima, Junzo Kuroda, and Yoshiharu Tsukamoto, *Made in Tokyo*, Kajima Institute Publishing, Tokyo, 2006.
[4] Akira Suzuki, *Do Android Crows Fly Over the Skies of an Electronic Tokyo? The Interactive Urban Landscape of Japan*, Architectural Association, London, 2001, p.48.
[5] Rem Koolhaas, 'Movement', in R. Koolhaas and H. U. Obrist, *Project Japan: Metabolism talks...*, Taschen, Köln, 2011, p.12.
[6] Nicola Russi, *Backgrounds, Quodlibert*, Macerata, 2018, p.127.

伝統と革新の
相対的関係

マウロ・ヴォルピアノ
（トリノ工科大学建築デザイン学部准教授［建築史］）

　欧州の学生、殊にイタリアの学生にとって、日本の都市の実地調査は、わくわくするような挑戦である。東京は発展を続ける素晴らしい大都市であり、過去に遡って辿ると、歴史のなかで明確にかたちづくられた西欧都市の形態や都市計画に幾分しか依拠していない。この都市ついて知識を深める挑戦であるのだから、なおのことそうである。文献に当たれば、東京について過度に単純化した解釈ができないことは明白である。槇文彦（『見えがくれする都市──江戸から東京へ』鹿島出版会、1980）や陣内秀信（『東京の空間人類学』ちくま学芸文庫、1992）などの日本の著者による文献だけでなく、イタリア人のフォスコ・マライーニのような歴史上の人物や、アンドレ・ソーレンセン、スティーブン・マンスフィールド、バリー・シェルトン等現代の著者による文献においても、東京の都市観察に伴う難しさが引き合いに出され、都市空間の複雑さおよびスケールその他の要素の多様性が主張されている。

　よって、今回のワークショップは学生のみならず教授陣にとっても、伝統と革新、アイデンティティと社会的な空間利用、建築遺産の歴史と実在における相対的関係に見慣れない特徴があるという現実に直面す

る機会となった。それでも、（未来を模索する大都市の一例でもある）東京が有形無形の歴史的特徴、記憶、伝統に富むことは一目瞭然であった。

　たとえば、近代や（ローマ帝国時代とまで言わないにしても）中世の時代性がそのまま建造物や街路に保存されている例があるイタリア都市と異なり、東京の特徴は存在と不在の複雑な相対的関係性にある。複雑で多中心的な都市のシステムにおいて不変の要素を把握するためには、街区と街路から成る都市計画、鉄骨造や鉄筋コンクリート造の高層建築と混在する伝統的な住居の存在、公共の建物や空間へと変貌した近世の上流階級の屋敷について、文献や地図に根気よく当たる必要がある。寺、歴史ある庭園など、主要なモニュメントについて吟味したり解釈する場合においても、この複雑なコンテクストを考慮しなくてはならない。

　東京では現代性と過去がそこはかとなく共存し、街で独特な活用のされ方をしているパブリックな空間やプライベートな空間に、モニュメント化されていない過去が存在している。このことは、訪れる者にとって最も興味深い知見のひとつである。

　それと同時に、外来のデザインやパラダ

A Challenging Dialectic between Tradition and Innovation

Mauro Volpiano

(Associate Professor of Architectural History, Department of Architecture and Design [DAD], Politecnico di Torino)

It is a fascinating challenge for European students – and in particular for Italian ones – to investigate the complexity of the Japanese city. Even more so when it is a matter of deepening the knowledge of Tokyo, an extraordinary metropolis in continuous transformation and whose characters can be only partially traced back to the morphologies and urban design of the western city as it has been outlined during history. The impossibility of any interpretative reductionism is evident in the scientific literature: the texts of Japanese authors and architects, such as Fumihiko Maki (*City with a Hidden Past*) or Hidenobu Jinnai (*Tokyo, a Spatial Anthropology*), but also those of Western scholars, already historicized like the Italian Fosco Maraini, or contemporaries like André Sorensen, Stephen Mansfield or Barrie Shelton, insist on the complexity, multiplicity and multiscalarity of urban space, invoking that difficult exercise of urban observation that Tokyo always requires.

This seminar was therefore an opportunity for all of us, students and teachers, to confront a reality where the dialectics between innovation and tradition, identity and social uses of space, history and materiality of built heritage take on very different characters from those to which we are familiar. Despite this, the Japanese city (and also a metropolis so strongly projected in the future as Tokyo is evidently no exception) immediately shows its abundance of historical characters, memories, traditions, which are expressed tangibly and intangibly.

However, unlike for example the Italian city – whose modern or medieval, if not even Roman, past is sometimes still fully preserved in buildings and streets – in Tokyo it is a complex dialectic of presence and absence that defines the character of the historical city. The urban plot of blocks and streets, the presence of traditional residential typologies that coexist with multi-storey buildings in steel and reinforced concrete, the ancient aristocratic estates transformed into public spaces and buildings, all this must be patiently investigated with documents and with maps to capture the permanencies of a complex and polycentric urban system. This urban context of great

イムが吸収され、独特な用いられ方がなされていることに都市の国際性が表れている。その例として、西欧の歴史主義建築の影響を受けた明治維新の頃の作品や、19世紀に欧州の都市計画モデルを応用してつくられた公園（今回の対象地区に囲まれた上野公園など）が挙げられる。

都市の歴史的研究方法論は西欧のパラダイムに立脚して久しいが、東京はその伝統的なアプローチの再考を促している。そして、私たちの同時代性を特徴づけている多様な都市の価値観や体験に光を当てる比較世界史の重要性を示唆している。この意味において、欧州、北米、アジアの学生および教授陣が集まった本ワークショップは、主催者側のおもてなしのおかげもあり、参加者一人ひとりにとって楽しくたいへん有意義な場であった。

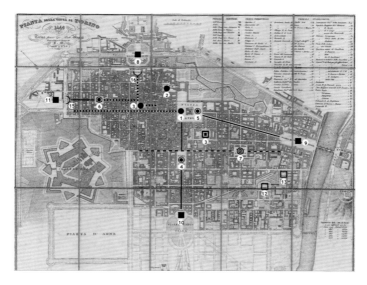

Turin: the main axes of the seventeenth and eighteenth-century urban expansions, the urban gates and the main functional and symbolic poles of the city are depicted on an 1848 printed map of the city. (The author's elaboration on the image. © Mauro Volpiano 2019.)／トリノ——17世紀と18世紀の都市拡張の主軸、都市の門、都市の機能的で象徴的な極は、1848年の地図に明示されている（筆者加工©マウロ・ヴォルピアノ、2019）

complexity cannot be neglected even in the interpretation and analysis of major monuments, such as temples or historic gardens.

This pervasive presence of the past that coexists with modernity, and which materializes, often without monumentalization, even in the peculiar use of public/private space by the local communities, is one of the most interesting lessons that Tokyo offers its visitors. At the same time, the internationality of the city has always been ev dent in the assimilation and very original reworking of non-native patterns and paradigms, such as those of Western historicist architecture in the years of Meiji modernization, or in the re-elaboration of European urban design models, such as the nineteenth-century public park, of which Ueno Koen, the center of gravity of our study area, is an example. Tokyo therefore urges us to reconsider the traditional approaches of urban historiography founded for many decades on almost exclusively Western paradigms, to suggest vice versa the relevance of a comparative global history capable of highlighting the multiplicity of urban values and experiences that characterize our contemporaneity. In this sense, the workshop among students and teachers of three different continents that we have experienced during the course of a busy week has been not only pleasant and enjoyable, thanks also to the wonderful hospitality we have received, but very useful and instructive for all of us.

トリノ再構築
——都市の未来、ハイブリッド空間

マルコ・サンタンジェロ
(トリノ工科大学都市地域研究計画学部准教授［地理］)

レクチャー①

　都市の文脈における課題とは何でしょうか。この講義では、トリノを題材に、建築、ランドスケープ、そして地域の3つのスケールを関連づけながら、都市改造の空間的展望を描くための枠組みを模索します。

　トリノは、山岳地にある小公国の首都でしたが、その後イタリア王国の首都に変わります。そして産業の中心地として栄えた後、典型的なポストフォーディズム（フォーディズムとは、労働効率化、生産力向上、消費喚起を組み込んだ経済成長モデルで、1950年代に普及した。ポストフォーディズムとは、その生産性鈍化とともに新たな労働編成が試みられたことを指す）のイタリア都市、現代アートの都、そしてバロック都市としての顔を持つに至りました。ローマ帝国時代を偲ばせる基盤の目のように組まれたトリノの市街地には、時代とともに変遷した都市の役割や姿が残されています。そして、16世紀以降に都市計画に組み込まれた王宮群が旧市街地に王冠を添えるように配されているのです。しかし今日トリノの街並みから公国首都時代の壮大な空間構想を読み取ることは容易ではありません。広域化した都市にそぐわない制度的枠組みのなかで、トリノの今後の展望が模索され、都市の新たな背景や課題に対応する構想が希求されています。

　トリノの風景をご覧ください [fig.1]。ここからは、都市の3つの象徴、心理状態、そしてハイブリッド化という観点からトリノについて話しましょう。

3つの象徴

　1つめの象徴は格子状の市街地です。旧市街地の輪郭にあった壁は、何世紀もの時を越えていまだにその一部が残されています。ローマ帝国時代につくられた格子状の街区も現存しています。2つめの象徴は旧市街地を囲む王宮群 [fig.2] が縁取って地図上に現れる王冠のようなライン、そして現在の市街地を囲む王宮群が縁取る同じく王冠のようなラインです。市街地の周りには格子状の街区に組み込まれなくてはならなかったサヴォイア公の邸宅（後の王宮群）が多数ありました。3つめの象徴は都市大改造のために配された3本の軸です。3本の軸について説明する前に、工業都市時代およびその時代終焉後のトリノの特徴について触れましょう。

Reframing Torino:
Hybrid Spaces of the Future City

Marco Santangelo

(Associate Professor of Geography, Interuniversity Department of Urban and Regional Studies and Planning [DIST] , Politecnico di Torino)

Lecture①

What challenges the urban fabric in a city? We will take Torino, or Turin, as an example to consider a spatial vision for transforming a city, linking different scales from the architectural to landscape to regional.

In its long history, Torino has been the former capital of a small mountain duchy and then – shortly – of a unified nation, a former industrial powerhouse, the Italian epitome of post-Fordism[1], the Baroque city, and the capital of contemporary art. All these phases, roles, and shapes have been kept together in a grid-like street structure recalling Torino's Roman origins. The street grid is then paired with royal palaces

planned since the 16th century, which crown the historic city center. However, the grand spatial vision of the duchy capital is hardly visible nowadays and the city struggles to imagine its future within institutional frameworks unsuited for an expanded urban area [fig.1]. To adapt to new scenarios and challenges, a vision for urban transformation is required. I would like to talk about Torino in terms of signs, psychology, and hybridity.

Three Signs of Torino

Let me introduce three signs representing Torino. The first sign is a grid. Torino has been developed according to a grid that maintains the original structure of the Roman settlement. The second sign is a crown: the royal palaces [fig.2], when seen from the above, dot the wider city area like a crown adorning the urban space. These palaces, which used to be ducal houses,

fig.1 Panoramic view of Torino／トリノの風景
Photo courtesy of Lorenzo Attardo／ロレンツォ・アッタルド撮影

fig.2 Royal Palace of Racconigi, one of the palaces of the royal crown around the former capital of the unified nation／かつてのイタリア王国の首都に王冠のように添えられた王宮群の一つ、ラッコニージ城
Photo courtesy of Lorenzo Attardo／ロレンツォ・アッタルド撮影

FIATの街

　過去に遡ると、トリノは自動車産業においてデトロイトと比較されるFIATの街でした。工業化によりフォーディズムの都市へと発展したトリノは、その後ポストフォーディズムの都市へと移行するのですが、まずはフィレンツェがイタリア最大の自動車工場になったところを想像してみてください。私見ですが、トリノはフィレンツェとデトロイトのフランケンシュタイン的な掛け合わせです。

　工業化は19世紀末に始まり、その後20世紀を通じてトリノの都市をかたちづくりました。トリノには建築的偉業といえるFIATのリンゴット工場 [fig.3] やミラフィオーリ工場などの巨大工場が建設されました。建物の大きさから推測されるように、トリノの人口は1953年から10年間でおよそ80万人から1.5倍に増加しました。イタリア北東部と南部から労働者が流入し、彼らのための家が必要になり、都市全体が産業を基盤としてつくられました。現在トリノの社会資本を構成するのは、当時の労働者を中心とする人々の子どもたちです。

抑圧の心理

　都市空間の利用に関する心理について簡単に触れましょう。抑制と抑圧を抱える都市には、変わった作用があります。抑制とは意識的行為であり、フラストレーションにつながりますが、抑圧とは無意識な心の働きであり、好ましくない影響を伴うと考えられます。トリノでは、自然の風景と文化的資質が抑圧されてきました。工業都市では産業が優先され、都市の資質に見過ごされたのです。fig.4はトリノを流れる4本の主要河川のひとつであるドーラ川です。この川はオフィスビルや工場建設のため、大幅に埋め立てられました。

　文化的資質の抑圧について例を挙げると、この写真の上部にあるのが、トリノ王宮・王室庭園の敷地内にある大聖堂ですが、王室庭園前の敷地は、当時駐車場として使われていた市街地中心の広場です。FIATの街らしい光景といえます。

　fig.5はポストフォーディズム期のトリノの街の心理的イメージを把握するのに役立ちます。ポストフォーディズム期（1970年代から2010年）には、何百万平米もの

fig.3　The former FIAT Lingotto Factory, now a multi-purpose building／
FIATリンゴット工場。現在は複合ビルとなっている
Photo courtesy of Lorenzo Attardo／ロレンツォ・アッタルド撮影

compose a system of royal residences. Before addressing the scratches, the industrial and post-industrial phases of Torino should be considered.

The One-Company Town: FIAT

Before the scratches were introduced, Torino was a one-company town, often compared with Detroit. It was a Fordist city, which then became a post-Fordist city. Imagine a city like Florence that became the biggest Italian automobile factory town. In my view, Torino is a Frankenstein mix of Florence and Detroit.

Industrialization began by the end of the 19th century and shaped the city throughout the 20th century. The city began to see vast factories like the FIAT Lingotto factory [fig.3], an architectural feat, and the FIAT Mirafiori complex. Considering the size of industrial buildings and the related workforce required, unsurprisingly, the city's original population of 800,000 grew by 50% in the decade following 1953. Workers came from the northeast and south of Italy, and housing was required. So, entire districts were built, weaving the city's social fabric with industry. Nowadays, the population of the city still consists primarily of the descendants of those workers and their families.

Psychology: Suppression and Repression

Let me talk briefly about psychology. There is a strange mechanism in Torino, a city of suppression or repression, namely, a conscious act that causes frustration or a subconscious damaging act. In Torino, there had been the astonishing suppression of environmental landscapes and cultural qualities, as it was an industrial city where the quality of the city was neglected in favour of manufacturing. For example, the Dora River [fig.4], one of the four major rivers in Torino, was covered for a large part to create space to build offices and factories on top.

Another kind of suppression was seen with the cultural heritage. For instance, right in the heart of the city, the cathedral, the Royal Gardens, and the Royal Palace all had parking lots built in front – which

fig.4 The Dora River when it was still covered by a massive concrete slab／コンクリートスラブで覆われていた当時のドーラ川
Photo courtesy of Marco Santangelo／筆者撮影

fig.5 Dismissed industrial areas in Torino／放棄された工業用地
Città di Torino, PRG 1995／トリノ市都市計画図、1995年

土地が放棄されましたが、放棄地の多くが環境汚染のために廃れた工業区域（ブラウンフィールド）でした。図の濃い灰色の地域が工業地域です。この時期に、トリノという都市のアイデンティティが実質的に再定義されることとなり、主に産業用の土地建物の再活用やコンバージョンが行われました。

3本の軸

1995年に認可された都市計画では、問題点、解決策の検討内容と解決策が示されています。産業の中心地としての時代の終焉を予期して都市の大改造が行われ、市街地をその周辺に対して開くために、格子状の市街地に切り込むように配された3本の軸が旧市街地と（広域の）王宮群の間に計画されました [fig.6]。これはトリノが社会経済的に新たな段階に入ったことを象徴する重要な要素でした。3本の軸が導入されたのは、かつての産業地区と川などの環境資源の再活用が必要であったためです。

3つの経路からトリノの都市の文脈を辿ることを想像してみてください。1本めの軸は図の左側の地域を通過する経路です。

2本めの軸は図の中心にある主要交通路です。12kmに及ぶ大工事の末、鉄道が地下鉄になり、その上に道路が整備されました。3本めの軸は図の右側にある自然と親しむための経路です。川の価値が再認識され、整備されました。こうしてトリノは変貌を遂げたのでした。

新たな都市の兆し、都市大改造案

2本めの軸の一環として整備された道路空間にはパブリック・アートが出現したほか、線路が路上に浮上する位置に高速鉄道の駅が設置され [fig.7]、道路沿いの工場跡地は展示スペースやコンサート会場として使われるようになりました。道路沿いの地域にはレンゾ・ピアノ設計による高層ビルなどの新たな建物が建てられました [fig.8]。再開発の波は旧工業地帯にも及びます。fig.9は自動車産業向けの製鉄工場の跡地です。この区画の汚染レベルは想像に難くありませんが、それはともかく、この跡地には公園が整備されました。工業跡地の再開発においては、工業用地であった当時の特徴を維持するか選択可能なのです。この公園一帯では、工業用建物の大部分が保存され、

fig.6　Three scratches／3本の軸
Author's elaboration on Dansero E. (1993), Dentro ai vuoti: Dismissione industriale e trasformazioni urbane a Torino, Cortina, Torino／E・ダンセロの書物の図版に筆者が描き加えた

is characteristic of Torino, a FIAT city producing cars.

A picture from the city masterplan helps us to understand the psychological view of the city during the post-Fordist phase from the 1970s to 2010 when millions of square meters of abandoned areas, mainly brownfields, were produced [fig.5]. The dark gray areas are the industrial areas. During the post-Fordist period, the "material" identity of the city was redefined, mainly through the reuse or conversion of former industrial spaces and buildings.

Three Scratches

The grand urban redesign that foresaw the end of the industrial capital era reconfigured the city: three metropolitan axes were designed to open Torino to its city-regional space [fig.6], scratching the grid and the crown to superimpose another important sign of a new social and economic phase. The metropolitan axes were introduced, as it was necessary to reuse former industrial areas and to rediscover precious environmental assets, specifically, the rivers.

Imagine going through the urban fabric of Torino in three different paths. One scratch is the path of the services on the left. Another scratch is the main transportation path, a result of a massive 12-km work to bury the railway underground and build a boulevard above ground. The other scratch is the leisure path on the right, recognizing the importance of the river. These paths drastically changed the city.

A Glimpse of a New City: Grand Redesign

On the new boulevard, public art has been installed, a new railway station for high-speed trains has been built where the railways are above ground [fig.7], and former factories along this main transportation path have been used for exhibitions and concerts. In this boulevard area, new buildings were built, including the Renzo Piano Skyscraper [fig.8]. Former industrial areas have also been

fig.7　Porta Susa station／ポルタ・スーザ駅
Photo courtesy of Lorenzo Attardo／ロレンツォ・アッタルド撮影

音楽祭や式典などに使われています。この地区の用途は多様で、大司教区の本部や最先端技術の研究開発を支えるインキュベーターとしても使われています。

　しかし、都市大改造案は、縦割り行政の硬直化により広域化したトリノ市の包括的な都市構想に基づいて市内の各自治体における空間の再活用が阻まれている問題や、一連の経済危機の煽りを受けてしまいます。3本の軸は導入されましたが、その後は広域化した都市とその展望にそぐわない局所的介入によって再開発が進められています。

ハイブリッド化

　都市が急速に変化するただなかで、建物や空間をトリノ市全体の戦略的な開発構想に組み込むために、「ハイブリッド化」することを検討したらよいのではないでしょうか。ハイブリッド化とは空間や建物の多用途化を意味します。既定の用途で空きスペースの活用法を縛ることなく、スペース活用の可能性を広げるアイデアです。空きスペース自体を問題視する向きがあり、空きスペースを解消するためだけに空間の用途が設定されてしまう事態を、ハイブリッド化という概念を取り入れて未然に防ぐことができます。そして、都市の未来像を考える前に、まずは都市の現状を考えることが可能になるのです。現状への対処こそが、じつはトリノという都市の未来像を描くための唯一の手段ではないか、と考えます。

［2018年7月21日講義より］

fig.8　Renzo Piano skyscraper／
レンゾ・ピアノ設計の高層ビル
Photo courtesy of Lorenzo Attardo／
ロレンツォ・アッタルド撮影

fig.9　Dora Park area／ドーラ公園地区
Photo courtesy of Lorenzo Attardo／ロレンツォ・アッタルド撮影

redeveloped. For example, in the premises that used to have the steel factories for automobiles – we can imagine the level of pollution in the area – a park has been created [fig.9]. In a former industrial area, we can retain or eliminate its industrial characters. In the park area, the large portion of the existing buildings were kept to be used for music festivals, ceremonies, etc. The area is also used for other functions, such as the new headquarters of the archbishop of the city and as an incubator for emerging technologies.

The grand urban redesign, however, has been impacted by a series of economic crises and rigid administrative boundaries hampering the spatial reuse under a wider territorial outlook for urban transformation. This resulted in a phase of acupunctural interventions, unfit for the expanded city and its broader vision.

Hybridity

To engage spaces and buildings with wider, strategic territorial development, we may consider the concept of "hybridity," as the city is rapidly changing. "Hybridity" here means adaptability to multiple spatial functions, which will unlock the potential of vacant spaces and buildings that are otherwise bound by predetermined functions. This hybrid approach can prevent assigning functions to vacant spaces and buildings merely to alleviate the abundant vacancy viewed as a problem by some. Moreover, this hybrid approach will allow us to consider Torino's current situation over planning how to represent the city. This hybrid approach focusing on the current situation may be the only way in which Torino can envision its future.

Marco Santangelo's summary of his lecture on July 21, 2018.

★1　Fordism refers to an economic growth regime spread in 1950s, involving increased labor efficiency, higher production capacity, and stimulation of consumption. The term post-Fordism means that, as productivity of the Fordist regime stagnated, workforce realignment was sought.

LA——対極的な部分が織り成す全体

ジョン・N・ボーン
（南カリフォルニア建築大学日本・中国国際プログラム・コーディネーター、ファカルティメンバー）

レクチャー②

都市を読む——LAと東京

ロサンゼルス（LA）と東京は、網状の都市に見られる性質をもっている点で共通しており、その複雑な都市構成には西欧の都市によくある同心円的な性質がないとロラン・バルトは述べています[1]。両都市では中心と周縁の関係が曖昧であり、この点で西欧都市の構造と異なります。槇文彦は、東京における曖昧な図－地の関係や、流動性、可変性を「雲」（「時計」の反対概念）に喩えています[2]。陣内秀信は、現代の東京に継承された江戸時代の土地のパターンや空間を辿っており[3]、北山恒らはピクセル化した街区のメタボリズムおよびコモン・スペースについて鋭い論述を展開しています[4]。海外の人のなかでは、ロラン・バルトが都市の「空虚の中心」である皇居、および街の中心を成すものの「［精神的に］空虚な中心」である駅について書き著しているほか、東京という都市の文脈については、バリー・シェルトンの書籍に、大通り沿いのコンクリートの高層建築の帯に低層の木造住宅地が挟まれた都市の構造（いわゆる「硬い殻と柔らかい黄身」）が詳述されています[5]。またジュリアン・ウォラルとエレズ・ゴラン・ソロモンらは現代の東京について考察し、ジェネリックな建物を都市構成要素のひとつとみなしています[6]。これらの解釈はLAの都市の文脈を読む際にある程度参考になりますが、21世紀のただなかで、都市の現状について考えるための新たな視点が必要です。つまり、人の移動と都市の体験に的を絞った視点です。これは人の移動の連続性に重点を置く視点であり、時間と空間のスケール、事実と虚構、過去、現在、未来における都市の諸相といった諸条件のなかで絶えず変化する人間の性質を組み合わせる視点でもあります。東京という都市を読む際にこの視点がどのように役立つか、私が携わっている（旧国道）66号線プロジェクトを題材に、具体的に述べたいと思います。

移動の体験

人類の歴史を通じて、人の移動の体験に重点を置く習わしがあります。いくつか例を挙げると、古代スペインのセント・ジェームス・ウェイの巡礼、江戸時代の中山道の旅、そして20世紀米国の66号線の旅があります。『カリクストゥス写本』（14世紀）や松尾芭蕉の『奥の細道』（1634）などの日記や著述、ギー・ドゥボールの「漂流」などの方法論、同じくドゥボールによる

The Idiosynthetic in Los Angeles

John N. Bohn
(Japan / China Studio Coordinator, Part-Time Faculty, SCI-Arc)

Lecture②

Reading the City:
Los Angeles and Tokyo

Roland Barthes has stated that both Los Angeles and Tokyo share qualities of reticulated cities, with a complex net-like organization that lacks the centrifugal qualities of typical Western cities[1]. Unlike Western cities, both cities operate without a clear center / periphery relationship. In the case of Tokyo, Fumihiko Maki used the metaphor of a cloud (in opposition to a clock) to describe its ambiguous figure / ground relationships, fluidity and instability[2]. Hidenoubu Jinnai traced Edo-era ground patterns and urban spaces to contemporary Tokyo's[3], and Koh Kitayama with co-authors wrote convincingly about a *Tokyo Metabolizing* of pixelizations and shared spaces[4]. Foreigners such as Roland Barthes have written about the "Center City-Empty Center"of Tokyo's Imperial Palace and the role of the train station "void" as a neighborhood center. Barrie Shelton effectively described Tokyo's urban fabric, with its tall concrete structures bordering the major streets and protecting the smaller wooden residential

neighborhoods behind – the so-called hard shell, soft yoke[5]. And Julian Worrall and Erez Solomon included "generic" structures in their consideration of Tokyo[6]. While some aspects of these readings of Tokyo could be applied to a reading of Los Angeles, perhaps moving well into the 21st century requires another way to consider the urban condition, one in which human movement and experience is our primary lens. This alternative might be structured more along the continuity of human movement through the city, synthesizing our own idiosyncratic oscillations between scales of time and space, facts and fictions as well as past, present and futures of the urban condition. To illustrate my point, I will take my Pathway project – Route 66 – as it passes through Los Angeles as an example of how that might be applied to looking into the city of Tokyo.

Experience of Human Movement

There is a strong tradition of emphasizing the experience of human movement along pathways that extends the length of human history. From pilgrimages along St. James

103

「Naked City（裸の都市）」(1957) などの描写、ジャック・ケルアックの小説『路上』(1957)、エド・ルシェのフォト・ブック『Every Building on the Sunset Strip（サンセット・ストリップ沿いの建物）』(1966) においては、移動の過程での体験が重視されてきました。

「空虚な」空間

　英国の建築批評家であるレイナー・バンハムは、LAの生態系を「浜辺の街」「山麓の丘陵地帯」「イドの平地」「高速道路」に4分類しています[7]。丘陵地帯の裾野に広がる盆地にはスプロール化が進み、多くのLA住民の職住の場となっていますが、「イドの平地」とはこの盆地を指しています。LA一の平凡な場所であると考えますが、同時にその平凡さが特徴的であるとみなされている場所です。ここで着目したいのが、イドの平地で読み取れるLAの特徴です。それは、LAではその空間の大部分をすきまの空間が占めており、すきまの空間どうしをつなぐ動線によって移動の体験がかたちづくられるということです。実際、街や多極的に位置する中心地ではなく、それらをつなぐ動線沿いのすきまの空間がLAという街を物語っているといえます。網状の都市とみなすだけではLAという都市の体験からかけ離れており、その体験はむしろ移動の連続です。絶え間なく、しかし変化があり、強烈で、押し流され、衝動に駆り立てられる、そういった移動の体験です。LAの都市の姿は、地図を見るだけでは把握できませんが、道を移動しながら理解できるのです。

ロサンゼルスの構造

　LAは、ほかの都市同様に事実と虚構から成り立っており、その最も顕著な姿をとっているといえるでしょう。人々の物語から不動産開発業者や娯楽産業の夢に至るまで、事実と虚構が共存しているのです。またLAは非常に多様な時間と空間のスケールから成り、異なるスケール同士を取りもつインフラ構築によって居住環境が構築されてきました。概してLAは美しくかつ平凡な都市として映ります。この都市では都市を構成する要素がさまざまな比重で混在しているという印象を受けます。LAを成立させる要因とは、異なるインフラ、スケール、人間の習性の類似性と連続性であり、この要因は、ほかの都市にも当てはまるといえます。この都市に人間の体験が動線沿いに織り込まれていきます。通りや脇道は景観を縫って進み、沿道に広がる居住の営みが蓄積されていくのです。LAでは対極的な要素がおそらく最も顕著に具現化されており、これらの要素は連続する空間構成に存在し、この空間内で事象は距離を置いてつながっています。人は主に、移動によって異なる対極的要素に関係性をつくっているのです。

　LAの構造をなんらかの基盤として考える際の与件が2つあります。1つめは時間と空間のスケールです。LAは地質系統や水資源などの自然の体系として生まれました。水資源の制御は街の発展に不可欠な要素であり、それは土地に対する地質学的なスケールでの介入でした。盆地内の河川の

Way (El Camino) in ancient Spain, to the Nakasendo in Edo-era Japan and all the way to 20th century America, Route 66. Experiences along paths of movement have been prioritized in journals and writings such as the *Codex Calixtinus* (14th cent), Matsuo Basho's *Narrow Road to the Deep North* (1694), methodologies such as Guy Debord's "*derive*," and representations of his "*Naked City*" (1957) in illustration, Jack Kerouac's *On the Road* (1957) in text and Ed Ruscha's "*Every Building on the Sunset Strip*" (1966) in photography.

"Empty" Spaces

In his famous book, the British architecture critic Reyner Banham described the four ecologies of Los Angeles as Surfurbia, Foothills, Plains of Id, Autopia[*7]. Located below the Foothills were the "Plains of Id" consisting of the flat areas of the basin where Los Angeles' infamous sprawl and most of its residents lived and worked. The "Plains of Id" are where I find to be the most common – the most unique in its banality as it is often perceived. Of particular interest here are those parts of Los Angeles found within the "Plains of Id." The lines of movement that connect these "empty" spaces in-between that make up most of Los Angeles are the pathways of human experience. It could be argued that the city of Los Angeles is more about these connective spaces found along pathways than any of its places or its multiple centers. Reticulated

is one way to describe the organization, but that is far from the contemporary human experience of Los Angeles. It is more a continuously moving experience, never broken, but certainly shifting, expanding, intensifying, drifting, and driven. It is not a city understood exclusively from above through maps, but understood while passing through the city and moving along the pathways found within.

The Los Angeles Anatomy

Like all cities, but perhaps here in its most virulent form, Los Angeles is a city of fictions as much as facts – from personal stories to developers' dreams and those of the entertainment industry. Los Angeles is also a city of vastly different scales of space and time that has relied on developing sophisticated infrastructure systems to mediate those scales for human habitation. Lastly, Los Angeles is a city of both great beauty and vast banality, often juxtaposed in common perception. In human experience, those conditions more commonly blend together, appearing simultaneously but with varying degrees of intensity. What makes Los Angeles possible, and to some extent every city possible, is the similarity and continuity of all of these disparate systems, scales and behaviors. Accumulated along paths of travel, the city is now braided with human experience. Pathways and their tributaries thread through the landscape and have collected artifacts of human habitation along

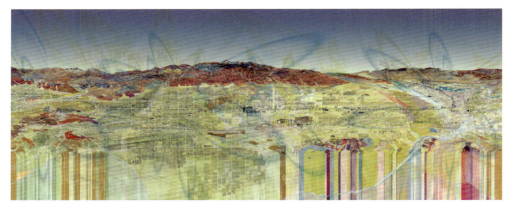

fig.1　Route 66 Pathway Project; Pomona to San Bernardino panel／66号線プロジェクト、ポモナーサンバーナディーノ間パネル
Quotation source: 2018 JBohn Associates／引用出典＝2018 JBohn Associates

流域、水路、そしてマンホールに落ちた鍵を拾おうとしている人を順番に想像してみてください。そして自然および地質学的なスケールから身近なヒューマン・スケールへの転換に目を向けてみてください。2つめの与件は各種媒体から得られた情報の記録とレイヤーの積層です。こうした記録の典型といえば地図や写真ですが、たとえば、土壌図、航空図やLAに水を供給する水道のパノラマ図などがあります。

66号線プロジェクト

　66号線プロジェクトはその沿道の空間を理解するための基盤をつくるべく、同空間での体験を再現する5カ年計画です [fig.1]。絵巻物のような構成で、サンタ・モニカーシカゴ間の沿道の空間に現れる事実や虚構を3次元で描き出します。多様なスケール、時間、空間を描き出すためにレイヤーを重ねつつ、AR技術を用いて音声や動画も組み込んでいます。第1対象地区は、イドの平地のポモナからその東にあるサンバーナディーノに抜ける区間です。バンハムのLA図に示されたLA周縁にあるすきまの空間どうしをつなぐこの道を中心に進めています。

　世界の情報が言葉や分類ではなく場所と時間を軸に体系化されていたとしたら、と想像してみてください。どうやって情報をまとめるのでしょう。この体系を用いてLAや東京といった都市について想像し、改めて表現したならば、都市を読み取る新たな方法が見つかるのではないでしょうか。たとえば、このケースでは、サン・ガブリエル山脈にUSGSの土壌データ、農務省林

these avenues of shared movement. Perhaps Los Angeles is the most extensive example of these extremes, existing simultaneously in a continuous system where instances remain simultaneously isolated and connected, their dissonances connected in many ways, but primarily by our movement among them.

When thinking about the Los Angeles anatomy represented as a productive platform, two conditions must be addressed. One is scales of time and space. Los Angeles began first as a geological formation and natural systems, including water. Controlling the movement of water was a geological-scale human intervention with the landscape, and vital to the city's development. Picture watersheds in the Los Angeles basin, aqueducts, and then somebody trying to fish keys from a manhole. Notice the shift in scale from the natural and geologic to human and immediate. The other is the registration and layering of information from various media. A typical example is maps and photographs, for example: a soils map, a pilot map, a panoramic map of the aqueduct bringing water to Los Angeles.

Route 66 Pathway Project

The Route 66 Pathway Project is a five-year-old project that endeavors to illustrate the human experience as it moves along Route 66 in order to provide a productive platform to understand these spaces [fig.1]. This series is structured much like a Japanese

emakimono, a continuous representation of the curated facts and fictions found along a pathway, extending the length of Route 66 from Santa Monica to Chicago. We model, or redraw three-dimensionally, these facts and fictions, layering and registering them to illustrate multiple scales, times and spaces, applying augmented reality technology to enable recordings and films to inhabit the pathway. The first section of the Route 66 Pathways Project covers the portion of Banham's "Plains of Id" along the route from the city of Pomona to the city of San Bernardino to its east. Its focus is on this portion of the pathway that connects the empty spaces at the periphery of Los Angeles County in Banham's diagrams of Los Angeles.

Imagine if all the world's information was not organized by word or category but by place and time. How might one put everything together? If we used this approach to imagine and re-represent cities, such as Los Angeles and Tokyo, we would discover a productive new way to look at the city. For example, in this case, we have mapped onto the San Gabriel Mountains the soil data from the USGS, fires mapped by the US Forest Service, the paths of flooding, housing developments, fault lines, flight patterns and watersheds.

Of course, as we move along a pathway, our perception layers and distorts, foregrounds

野部の山火事マップ、洪水の経路、住宅の発展、断層線、飛行航路のパターン、水路などをレイヤーとして重ねています。

　もちろん、私たちの知覚は、移動中に見えるものやそこから読み取ったことをレイヤー状に重ねたり、変化させたり、特定の部分にフォーカスしたり、思い出したり、思い巡らしたりします。たとえば、66号線を車で走っていて、サン・ガブリエル山脈を見ながら「なんて美しいのだろう……」と思い、右折しつつ故障車を視界に認め、さらに少し進むと見えてくる公園で自転車に乗ったときのことを思い出し、あの有名な建物で撮影された映画を思い出す、といったように。道は東から西へ真直ぐ延びています。もとを辿れば、アメリカ・インディアンが通った道でした。やがて小さな丘が見えてきます。直線続きの場所で出くわす変わった地形。……この地形はじつは大きな断層なのだとわかります。断層から泉が湧き、この地域で水が利用できるようになり、動物が生息し始め、人々もこの地に定住を始め、やがてこの土地の奪い合いが起こったことがわかります。この土地には殺人、戦い、耐え忍んだり衝突した話が残っていますが、地元の人々が語らない限り、過去に埋もれたままとなります。こうした人々の物語や、この道沿いで発展した住まうという営みの蓄積を掘り起こすことで、この都市に輝く素晴らしい瞬間と出会えます。

　世界が道に沿って体系化されていたとしたら、と想像してみてください。世界が場所の集合に限らないとするならば、それはある場所とほかの場所の間にある事象の体験の集積です。平凡な事象、美しい事象、あるいはとても身近な事象です。世界の見え方はどのように変わるでしょうか。

　66号線プロジェクトは、沿道にあるインフラ、構造物、特色、文化、歴史、そして体験を重ね合わせてまとめることを目的としています。66号線に刻まれた諸相にこそ、LAという都市の体験の全容、すなわち事実、虚構、歴史、未来までもが表れているのです。東京同様に、LAの諸相、すなわち対極的な諸相は、面的に織り成されて全体を成しています。東京は縦の空間利用が活発で、都市としてLAの一歩先を行く点で興味深いですが、東京の道と都市としての未来を探るうえで、66号線の研究法が役立つのではないかと考えます。

［2018年7月21日講義より］

★1　ロラン・バルト『表徴の帝国』(宗左近訳、ちくま学芸文庫、1996)
★2　Fumihiko Maki, Serge Salat and Françoise Labbé, "Fumihiko Maki: An Aesthetic of Fragmentation", 1988. (Barrie Shelton, Learning from the Japanese City: West Meets East in Urban Design, Taylor & Francis, 1999に所収)
★3　Hidenobu Jinnai, "Tokyo, Then and Now: Keys to Japanese Urban Design", 1987. (Barrie Shelton, Ibid. に所収)
★4　北山恒＋塚本由晴＋西沢立衛『TOKYO METABOLIZING』(TOTO出版、2010)
★5　Barrie Shelton, Learning from the Japanese City: West Meets East in Urban Design, Taylor & Francis, 1999.
★6　ジュリアン・ウォラル＋エレズ・ゴラニ・ソロモン＋ジョシュア・リーバーマン『英文版 東京現代建築ガイド──Twenty-first Century Tokyo: A Guide to Contemporary Architecture』(講談社インターナショナル、2010)
★7　Reyner Banham, Los Angeles: The Architecture of Four Ecologies, University of California Press, 2009.

and backgrounds, remembers and speculates as much as it looks or reads. For example, as you drive along Route 66, here you might look at the San Gabriel Mountains and reflect on how beautiful they are. Then, as you turn to your right, you see a car incident. You drive a little further and remember the time you rode your bike in that park, and remember a film clip filmed in that famous building.

The Route 66 pathway moves through this portion of Los Angeles in a straight line from east to west. It evolved from the trails of native Americans to roads for automobiles. Here is this small bump, an exception to the dominant linear direction of this pathway. Why? It turns out th s geological anomaly is a fault, a variatior that created a spring which provided access to water here, leading to the settling of animals, then people, and eventually a contested piece of property. The area houses stories of murders, battles, bears, and conflict that are unheard unless told by locals. Engaging with these personal stories and the features of human habitation and evolution along this pathway illustrates these fantastic moments that are found in Los Angeles.

Imagine if representations of the world were organized along its pathways. It is not necessarily just a world of places, but perhaps more a world of human experience of the context in-between – the banal,

beautiful, and the intimate. How might this change our understanding of the world?

The Route 66 Pathways Project collects, layers, and curates the various systems, structures, singularities, cultures, histories, and experiences found along the route. It is this condition that expresses the human experience of the urban fabric of Los Angeles in its entirety: facts, fictions, histories, features, and even futures. Los Angeles shares with the city of Tokyo this horizontal synthesis of the idiosyncratic – the *idiosynthetic*. While interestingly Tokyo goes one step further, operating into the vertical much more extensively, this method of studying this great pathway through Los Angeles might have some value for understanding Tokyo's own pathways and its urban future.

John N. Bohn's summary of his lecture on July 21, 2018.

★1   Roland Barthes, *Empire of Signs.*
★2   Fumihiko Maki, Serge Salat, and Françoise Labbé, *Fumihiko Maki: An Aesthetic of Fragmentation*, 1988, cited in B. Shelton, *Learning from the Japanese City: West Meets East in Urban Design*, 1999.
★3   Hidenobu Jinnai, *Tokyo, Then and Now: Keys to Japanese Urban Design*, 1987, cited in B. Shelton, *Ibid.*
★4   Koh Kitayama, Yoshiharu Tsukamoto, and Ryue Nishizawa, *Tokyo Metabolizing*, 2010.
★5   Barrie Shelton, *Learning from the Japanese City: West Meets East in Urban Design*, 1999.
★6   Julian Worrall, Erez Golani Solomon, and Joshua Lieberman, *Twenty-first Century Tokyo: A Guide to Contemporary Architecture*, Kodansha International Ltd., 2010.
★7   Reyner Banham, *Los Angeles: The Architecture of Four Ecologies*, University of California Press, 2009.

「江戸東京」という都市のコンセプト

北山 恒
〔法政大学デザイン工学部建築学科教授〕

レクチャー③

ラテンヨーロッパ／
アングロアメリカ／ファーイースト

　今回のワークショップは、トリノ工科大学と南カルフォルニア建築大学という、ヨーロッパと北アメリカからの大学が参加しています。ラテンヨーロッパと、アングロアメリカは、同じ欧米というキリスト教を背景とした文明圏にありますが、少し異なる都市観、または、文明観があると考えます。そこに、ファーイーストの日本が参加することで、文明の三角関係が生まれるのではと期待しています。

　12世紀から始まるヨーロッパ文明の世界への伝播はヨーロッパを地図の中心として、時間軸のなかで空間的拡張として表すことができます。1868年の明治維新は、日本という国家のシステムを、ヨーロッパ文明の社会システムに切り替えた切断面です。明治維新は、ジョルジュ・オスマンによるパリの大改造が終了するのとほぼ同じタイミングですが、パリという都市空間は、政治的な図式が明示されています。放射状になった道路パターンは監視を要する都市のように読める。ある意味ではパブリックという概念が空間で示されている都市です

[fig.1（パリ）]。パブリック概念が都市内で重要な意味をもつのは、ラテンヨーロッパが示す近代都市の姿です。明治維新の3年後、1871年にシカゴの都市中心部を焼き尽くすシカゴ大火が起きます。それを契機に北米に経済活動を中心とする「現代都市」という類型（タイポロジー）が登場しました。シカゴに引き続き20世紀初頭に急激につくられたニューヨークは「現代都市」を代表する資本主義がつくる都市であり、この経済活動を中心にした都市が「現代都市」タイポロジーです [fig.1（ニューヨーク）]。これは北米を中心としたアングロ・アメリカの思想を背景とした現代都市の姿です。

　ワークショップの対象地は上野公園の周囲を取り巻く6地区としましたが、そこでは江戸時代に寛永寺の境内の周囲にあった、町人地、武家屋敷を継承した江戸のコンテクストが読み取れます。ファーイーストという世界の果てにあった日本は、明治維新まではヨーロッパ文明を選択的に取り入れることができていたものの、明治維新のときに、一気に社会システムをヨーロッパ文明のシステムに切り替えています。江戸東京という都市を研究対象とするときは、ヨーロッパ文明を相対化することが必要です

The Urban Concept of Edo-Tokyo

Koh Kitayama
(Professor, Department of Architecture, Faculty of Engineering and Design, Hosei University)

Lecture ③

Latin Europe / Anglo-America / the Far East

The participants for this workshop include members from Politecnico di Torino and the Southern California Institute of Architecture. Latin Europe and Anglo-America belong to the Western civilization zone based on Christianity; yet, each has their unique urban and cultural perspectives. Blending these participants with those from Hosei University in Japan, this workshop will create an opportunity for the three civilizations to meet.

The spread of European civilization since the 12th century can be illustrated as a chronological spatial expansion on the Europe-centered world map. The timing of the 1868 Meiji Restoration in Japan was close to when Baron Georges-Eugène Haussmann completed transforming Paris. The urban space of Paris represents a political hierarchy, where its radial road

fig.1
パリ／Paris
画像 ©2019 Aerodata International Surveys, DigitalGlobe, The GeoInformation Group | InterAtlas　地図データ ©2019 Google
ニューヨーク／New York
画像 ©2019 Bluesky, DigitalGlobe, Sanborn, USDA Farm Service Agency　地図データ ©2019 Google
東京／Tokyo
画像 ©2019 Digital Earth Technology, DigitalGlobe, The GeoInformation Group　地図データ ©2019 Google

引用出典＝Googleマップ／Quotation source: Google Map

111

fig.2 ノリのパルマ、東京／Nolli's map of Parma; Aerial view of Tokyo
引用出典＝コーリン・ロウ＋フレッド・コッター『コラージュ・シティ』（鹿島出版会、2009）（パルマ）、法政大学北山研究室（東京）／
Quotation source: Colin Rowe and Fred Koetter, *Collage City*, Kajima Institute Publishing, 2009 (Parma); Kitayama Lab., Hosei University (Tokyo)

[fig.1（東京）]。そこで、江戸東京という都市の現代的な視点を探ってみたいと思います。

江戸東京の切断面

　明治維新は日本という国家のシステムを、産業化を中心とした社会システムに切り替えた切断面であり、この切断面によって鏡面のように江戸と東京を比較することもできます。江戸は人口の停滞した安定した定常型社会でしたが、明治維新後の近代化（産業社会化）によって、急激に人口が増加します。現在、日本は世界のなかでも先験的に人口が減少する国となっていますが、そのピークを打った人口動態を見ると「近代」という時代の異常さは認識できます。さまざまな科学技術が世界中の驚くべき人口拡大と、都市への集中を支えてきました。この時期はモダニズムの時代と一致します。20世紀末、都市型社会に移行した後、先進国ではその都市での出生率が下がっており、21世紀は急激な人口減少の時代となるのです。

　江戸期の人口は停滞しており、その推定人口は3000万ほどとされています。人間が食べる食物をつくる農業を基盤とする社会は資本の余剰を生まないので、定常型社会となります。「近代」という社会では進歩と拡大拡張を求め、産業化することによって剰余価値を生み、さらに効率よく生産を求めるために都市に人口が集積します。江戸は近代社会とは異なる自然と人工物が共生する集合形式で組み立てられた、冗長な空間を持つ都市なのでした。明治維新直後に地租改正条例（1873）が出され、土地の所有権が法的に認められ、市場経済を基盤とする近代都市の空間システムに変換されていきます。市中の面積の65％ほどを占めていた武家屋敷などの大きな囲い地が開発者の所有となり、それが細分化され区分所有されていきました。この土地所有の形式は継続されており、現代の東京の姿を決定づけているのです。

　連続壁体で構成されるヨーロッパの都市とは異なり、東京は「細粒都市」と形容される独立した粒の集合でできています。連続壁体で構成される都市は変化しません。18世紀半ばにつくられたノリの地図は現在の都市空間を示しています[fig.2]。

　独立した粒の集合でつくられた都市は容易に変化します。東京の建物の平均寿命は

pattern reflects the then government's intention to conduct public surveillance [fig.1(Paris)]. In a sense, the urban space of Paris represents the notion of "public" – an essential concept to modern Latin European cities. Three years after the Meiji Restoration, the Chicago Fire of 1871 burned down the center of the American city. Consequently, North America saw the formation of a new city typology: the contemporary urban city centering on economic activities. In the beginning of the 20th century, Chicago followed New York in rapidly evolving into a typical contemporary city shaped by capitalism. Today, the urban typology of contemporary cities refers to those centering on economic activities; the backdrop to these contemporary cities is the Anglo-American ideology [fig.1(New York)].

In this workshop's six target zones around Ueno Park, we can read the urban context of the Edo period when the districts of tradesmen, artisans, and samurai surrounded the Kan'ei-ji Temple precincts. Located in the Far East, Japan had been able to selectively adopt customs from European civilization until the 1868 Meiji Restoration when Japan's social system switched to a more European style of civilization. Therefore, studying the city of Edo-Tokyo requires us to compare it to European civilization. This lecture aims to explore contemporary perspectives for urban studies of Edo-Tokyo [fig.1(Tokyo)].

The Edo -Tokyo Demarcation

After the Meiji Restoration, Japan's national system became oriented to industrialization [and the city of Edo was renamed to the city of Tokyo]. Therefore, the Meiji Restoration is a reference point for comparing the cities of Edo and Tokyo. The city of Edo had a steady-state society with a stable population, but modernization after the Meiji Restoration industrialized society and drove a population surge. Today, Japan is among the nations expected to see a population decrease. Having peaked, the population dynamics suggest that the modern population jump was abnormal. Various scientific technologies have helped dramatic population expansion and urban migration worldwide. The period of the population expansion coincided with the age of modernism. At the end of the 20th century, urban birth rates dropped in developed countries where society had transitioned to an urbanized society. In the 21st century, these nations' urban populations are said to plunge.

During the Edo period, Japan's population stagnated around an estimated thirty million. A society based on agriculture, the industry producing food, generally does not generate surplus capital and becomes a steady-state society. In contrast, modern society generates surplus value through industrialization, pursuing progress and

113

fig.3　第12回ヴェネチア・ビエンナーレ国際建築展〈2010〉展示風景／
Venice Architecture Biennale 2010
撮影＝architecture WORKSHOP／Photo courtesy of architecture WORKSHOP

26年ほどです。「江戸東京」の都市形成は、起伏に富んだ豊かな丘陵地の地形を基盤としており、グリッドや同心円という人為的区式で街をつくるのではなく、自然地形を読み込み、それに応答して尾根道や谷道といった道が通され、街割りがつくられているのです。土地の敷地割りや道路パターンは、上物が変更されても継続されます。東京の都市の文脈は建物という実体ではなく、それを支えている地形、そしてそれに応答してつくられた基盤構造にあるのです。

「江戸東京」の都市組織

　東京では建物というソリッドを見ても時間のなかで継続する類型（タイポロジー）を読み取るのは困難ですが、ソリッドとソリッドの間に生まれるヴォイドに注目すると、地割りや道路パターンがつくる江戸東京の都市構造を読み取ることができます。建築の類型ではなくヴォイドのタイポロジーとして地図上で調べてみると、江戸東京という時間のなかで継続する都市構造が読み取れます。それを、面的ヴォイド、線形ヴォイド（p.55参照）、粒状ヴォイドと類型化します。面的ヴォイドとは、江戸の武家屋敷が公的な施設（公園、学校や官庁など）に変換したものと、寺社地境内として江戸時代から継続するもので、江戸期から数百年続く大きな空地が読み取れます。都市組織でいえばモニュメントに当たる役割をしています。線形ヴォイドは、地形地理と密接に関係しており、人が往来する商店街や道路、自然地形である水路やその暗渠、崖線緑地などの線形を読み取ることができます。この線形ヴォイドは生活に近接して存在するためコミュニティと深く関係しており、粒状のヴォイドは空き家・空き地が生まれやすい東京の都市構造を表現しています。面的ヴォイドや線形ヴォイドは人間の生命スパンを越えて継続されているのですが、この粒状の都市要素は明滅するように変化するのです。2010年の第12回ヴェネチア・ビエンナーレ国際建築展で、「TOKYO METABOLIZING」というタイトルをつけて、この絶え間なく生成変化を続ける粒状の都市組織を対象としたプレゼンテーションを行いました。この生成変化し続ける粒状の都市要素で埋め尽くされる東京の木造密集市街地にこそ、この都市の未来をつくる可能性があるとするのです [fig.3]。

expansion. As modern society seeks production efficiency, urban areas become more populated. Unlike modern society, the city of Edo afforded a generous urban space comprised of a collective form where nature coexisted with artificial structures. In 1873, following the Meiji Restoration, the land tax reform regulations were issued to legitimate land ownership. The city of Edo was then transformed to a city with a modern, urban spatial system grounded on a market economy. About 65% of the urban land constituted large lots, including samurai precincts, and these lots were acquired by developers to be subsequently divided into plots for sectional ownership. Continuing into the present day, this land ownership practice has dictated the contemporary urban space of Tokyo.

Unlike European cities shaped by a series of walls, Tokyo is shaped by an assemblage of independent buildings that look like grains on a map and can be described as a "pixel city" [fig.2]. A city comprised of independent grains easily changes, and in Tokyo, houses are scrapped on average only 26 years after construction. In contrast, cities shaped by a series of walls remain the same, as can be seen from a famous map of Nolli drawn in the mid-18th century, and yet still illustrates today's urban space.

Edo-Tokyo is a city developed on the hilly landform with versatile surface features. The city is not organized according to a planned schema, such as a grid or concentric circles. Instead, its zoning and roads (including ridgeways and valley ways) respond to the reading of the natural landform. The zoning and road patterns continue to exist even though buildings are replaced. Hence, the urban context of Tokyo lies not in buildings, but in the landform and the substrate responding to the landform.

The Urban Fabric of Edo-Tokyo

In Tokyo, lasting architectural typologies can be unclear even when we study buildings, or solids. Yet, when we look at voids between solids, or void patterns on maps, we can see the urban structure shaped by zoning and road patterns, which continues to exist from the Edo period. There are areal, linear (see p.55), and granular voids. Areal voids can be categorized into two: the lands of Edo-period samurai residences that are now used as parks, schools, government offices, or other public areas; and Edo-period temple or shrine precincts. From the perspective of urban fabric theory, these areal voids remaining for hundreds of years function as monuments. Linear voids are intertwined with topography, for example: shopping streets and roads, where people pass; naturally formed waterways and their culverts; and green ways along cliff lines. Embedded in communities, these linear voids are found in daily life. Granular voids reflect the urban structure where vacant

fig.4 「続・TOKYO METABOLIZING」展（2018）／Sequel to the Tokyo Metabolizing exhibition　撮影 = YUKAI／Photo courtesy of YUKAI

ヴォイド・インフラ

木造密集市街地の街区の最も奥にある未接道宅地を、共有のオープンスペースとするという研究を大学で行っています。このオープンスペースに接続して周囲の建物の共同建替えを誘導するものです。このオープンスペースを「ヴォイド・インフラ」と名づけました。周囲の住居はこの「ヴォイド・インフラ」を重集合として共同化し、さらに家と家のすきまや細街路とつないで歩行ネットワークを形成します。この「ヴォイド・インフラ」は耐火耐震壁という工作物で囲うのですが、建築物を建てることで問題を解決するのではなく、建物を除去することで地域のなかにポテンシャルを生もうとすることが、これまでの建築の概念とは異なっています。東京とは小さなグレイン（粒）の集合体でできた都市であり、その小さなグレインが自己都合で生成変化する不思議な集合体です。その都市のもつ特性をそのままに、未来の住人を柔らかい共同体という泡に包む空間組織が用意できないかという研究です [fig.4]。

「ヴォイド・インフラ」と細街路のネットワークは、そこを使う人々の出会いの場所をつくり、顔見知りの関係＝あいさつをする関係をつくります。それが、都市のなかにムラのような地域社会をつくるのではないでしょうか。都市はムラの泡が重なりながら寄せ集まっているというイメージです。そのムラが寺社地という大きなオープンスペースに接続すれば、生命スパンを超えたコミュニティスケールを覚醒させることができるのかもしれません。寺社地は、江戸から継続する数百年の間、都市内で巨大なオープンスペースとして存在しており、西欧のモニュメントのように都市の記憶となっているのです。この寺社の存在によって人々が集まって住まう根拠（たとえば祭り）を示すことが見出せれば、私たちの生活するこの〈都市〉を、江戸のムラから帰納される「非都市」に変換できるのかもしれません。

［2018年7月21日講義より］

houses and lots tend to be generated. Unlike the areal and linear voids that remain beyond a human lifespan, these granular voids change. When we fast-forward through a long-term view of these granular voids, they change like flashing lights. In the 2010 Venice Architecture Biennale, I gave a presentation entitled "Tokyo Metabolizing," focusing on Tokyo's urban structure where "grains" continue to change. In Tokyo, such grains constitute dense areas of wooden houses. Hence, in these areas lies the potential for building the future of Tokyo [fig.3].

Void Infrastructure

In my laboratory, we have been studying how to develop a common, unroofed space on a residential site tucked away from a legally sufficient frontage road and deep into a dense area of wooden houses, by replacing decrepit buildings with this common space. By creating this common space, "void infrastructure," we aim to promote the joint reconstruction of surrounding buildings. The void infrastructure abuts surrounding houses and shapes a footpath network reaching between houses and to alleys. In addition, the void infrastructure is surrounded by fire- and quake-proof walls. This infrastructure project aims to generate local potential by removing buildings, unlike the conventional architectural approach of adding buildings as solutions. Tokyo's urban fabric consists of an assemblage of grains, which are changed

at the landowner's discretion. While retaining this particularity, this project endeavors to introduce a spatial structure that will incorporate prospective residents of the project area into a loosely knit community [fig.4].

The void infrastructure and its footpath network will provide the opportunity for locals to meet and become neighbors who greet each other regularly when they meet. This will form an urban neighborhood similar to a village community. Our vision is that when this neighborhood concept is widely applied, these circles of neighbors partially overlapping with each other will comprise a city. Connecting these neighborhood circles to a larger open space of temple or shrine precincts may shape a community that can last longer than a human lifespan. Temple or shrine precincts – constituting large open areas, standing just like Western monuments, as part of living memories of the city of Tokyo from the Edo period – will potentially enable individuals to find meaning in gathering to live, or reasons for living together, such as festivals. We will then be able to build a "de-urban" Tokyo, the city's future founded on the essence of society in Edo-era villages.

Koh Kitayama's summary of his lecture on July 21, 2018.

新たな都市国家
——都会／田舎

ヘルナン・ディアス・アロンソ
（南カリフォルニア建築大学ディレクター兼CEO）

都市に見られる事象のなかでも、都市国家の時代への逆戻り現象は注目に値します。都市における政治の関係性について考えると、都会に住む人と田舎住まいの人との隔たりが顕在化してきています。もはや右派か左派かだけではありません。イデオロギーの変化が顕著に出始めており、所得格差にも同じことがいえます。たとえば、ロサンゼルス（LA）に住むある人の所得は東京やトリノに住んでいる誰かよりも多いかもしれず、これも政治に影響するのです。

密度＝効率

3都市についてのプレゼンから、各都市が都市全域を活性化するうえで直面している課題がわかりました。

トリノはイタリア内外において独特な都市です。またトリノ、ローマ、フィレンツェは活気があり、変化する都市である一方で、歴史の重みや密度という観点で欧州からの重圧にさらされている都市でもあります。グローバル化、エコロジー、持続可能性というテーマに取り組む際に、（都市の）密度が効率化をもたらすことに留意しなくてはなりません。

北山教授が述べられているように、東京は高密度に建物で覆われていますが、縦の空間はそこまで利用されていません。ジョン・N・ボーン教授が触れたように、都市、LAを読み解くには現象学的アプローチから都市化に迫る必要があります。過去半世紀、LAにおける興味深い事実は、人口流入の多い街のひとつであるということです。商業地区は再生され、都市、LAは急速に変化しています。小都市（micro cities）との関係性という点において、今後のLAの役割に関心をもっています。

アイデンティティの維持

今日都市は、欧州の都市に見られるように、アイデンティティの維持という課題に直面しています。欧州では人口増加が続いていますが、LAでは過去50年に極端な人口膨張が見られました。こうした人口変動により、都市のアイデンティティをどのように維持するかという課題が生じます。欧

New City States: Urban and Rural

Hernan Diaz Alonso
(SCI-Arc Director / CEO)

Political Division: Urban Inhabitants vs. Rural Inhabitants

One notable phenomenon of cities is that, clearly, we are back to the city-state era again. When we think of the relations of politics of the city, a contrast between urban and rural inhabitants is becoming more pronounced. No longer political division is only between right wing and left wing. We have begun to see a very distinct ideological change. Division between income levels is another; for example, an inhabitant of Los Angeles (LA) might have more income than an inhabitant of Tokyo or Torino and that too has an effect on their politics.

Density is Efficiency

From the presentations, we could see how the three cities are struggling to power the entire city.

Torino is a very unique city both in the Italian and global contexts. In addition, Torino,

Rome, and Florence are among the cities that are alive and changing while being subject to the weight of history and, in terms of density, that of Europe. When we address globalization, ecology, and sustainability, one point to consider is that density increases efficiency.

As described by Professor Kitayama, Tokyo is dense while not very vertically built-up yet. LA has another phenomenon. As narrated by John, reading the LA phenomenon requires a phenomenological approach to urbanism. One interesting aspect about LA in the last fifty years is that it is one of the cities more people migrated to. Its downtown has been revitalized and the city is rapidly changing. Therefore, I am curious to see the city's future role in relation to the micro city.

Identity Struggle

Today, as seen in European cities, there is another struggle: maintaining identity.

州の都市においては、都市のアイデンティティが歴史と深く結びついていますが、LAは都市スケールで本格的に発展したのが第二次大戦後になってからで、まだ若い都市であり、定まったアイデンティティのない都市に近いのです。LAの印象は、住民の間でも十人十色でしょう。

都市のもつ政治的な力

　近未来の都市の役割を理解するために、あるいは地球規模の課題に対する進歩的な取り組みに携わるとしたら、東京、LA、トリノの3都市をどの角度から見たらよいでしょうか。都市が完全に都市国家へと変貌したかどうかは、興味深い点です。都市は地理的に帰属する地域・国家からより大きな自由を獲得し、独自性を高めています。米国のトランプ現象について言えば、カリフォルニア州の都市等は、連邦政府を無視して商業、外交における独自の展望を描き始めています。米国ではこうした独自路線が出始めています。私たちは、都市の持つ権利が変わる様子を目の当たりにしている

のです。

　グローバル化とは、どのように解釈されようとも、大都市における共通性に通じる現象です。トリノ、ロンドン、サン・パウロ。どこであれ都会の生活には共通性があります。各都市に根づく伝統は固有のままですが、東京都民と世界の大都市の住民には、都民と日本の小都市の住民よりも多くの共通項が見られるのではないでしょうか。

　今後、都市にまつわる対話において、都市がもつ政治的な力、そしてその効果を高めるために、その政治力をどのように活用したらよいかが焦点となります。エコロジーや持続可能性といった地球規模のテーマに取り組むうえで、都市がもつ政治的な力に可能性があると考えます。

Europe is seeing a population increase; LA has seen an excessive population surge over the last five decades. These population dynamics can entail the problem of how to retain the city's identity. European cities have entrenched historical issues intertwined with their identities. In contrast, LA is closer to a city of non-identity, as the city is young, only being developed on a fully urbanized scale after WWII. Each LA resident has a unique view of their own LA.

Political Power of the City

To understand the significant future roles of the cities and if we want to participate in a progressive agenda, how do we approach the subjects of Tokyo, Torino, and LA? I am curious to see whether cities have fully become city-states. Cities have been increasingly becoming autonomous, freer from their countries or regions. Take the Trump phenomenon in America. Cities in California, in particular, began to build their own commercial, diplomatic vision, ignoring the federal government. We have begun to see this phenomenon in America. In these moments we live, we are observing changes in the rights of the city.

Globalization, however viewed, is a phenomenon leading to commonality in metropolises. Whether in Torino, London, São Paulo, you will find commonalities in the urban way of life. Residents of Tokyo probably have much more in common with these urbanites than with small-town inhabitants in Japan even while each metropolis maintains its distinctive traditions.

Going forward, the focus of conversations on the city will be the political power of the city and how we can engage that power to increase its beneficial effects. I can see hope in the political power of cities in tackling global challenges on ecology and sustainability.

東京の現状と今後への期待

陣内秀信
(法政大学江戸東京研究センター特任教授)

都市計画への市民参加

マルコ・サンタンジェロ先生のトリノに関する説明を聞きながら、[3本の軸導入における]都市計画上の戦略を理解することができました。トリノははっきりとした戦略的な展望をもっていたはずです。計画策定には市民も参加したのかもしれません。都市再開発の典型的な手法が、市民を巻き込むことです。

日本ではしかし、これは稀です。東京都の再開発は、1980年代後半までは東京都主導で進められていました。その例が東京湾南部の中心地の開発です。しかし、90年代に入る頃には、ディベロッパー等民間組織が開発の主導権を握り始め、丸の内、六本木、渋谷、日本橋といった地区で再開発が行われました。開発事業において、都は主役を降板し、サポーター役になりました。

ボストンの港湾当局を訪れた際に、自治体の都市計画家が戦略的な街の展望を説明してくれました。一方、東京都の都市整備局には、明確な展望がありません。都市開発においては、自治体と市民双方の視点を取り入れることがより重要になってきています。また(大規模な取り組みに限らず)個々の草の根の活動の連携も不可欠です。日本にはヒューマン・スケールの面白い建築づくりに率先して取り組んでいる建築家が大勢います。

日本における草の根の活動

銀座では興味深い取り組みが行われています。ビル所有者からなる組合があり、建設業者相手に闘い、建築物の高さを56mまでとする規制を勝ち取ったのです。今日銀座における建設計画は、街の委員会によって検討、評価されます。

一方、谷中には風通しの良いコミュニティがあります。谷中のコミュニティづくりに貢献している活動の例として、森まゆみさんらによる地域雑誌の発行があります。地元の人の連携によって、安心して住める街がつくられています。

トップダウン、ボトムアップの関係性

トップダウン、ボトムアップの関係性とトップダウン式における官と民(ディベロ

The City of Edo - Tokyo and My Hope for its Future

Hidenobu Jinnai

(Professor, Hosei University Research Center for Edo - Tokyo Studies)

Citizens' Participation to Urban Planning

Listening to the presentation on Torino by Prof. Marco Santangelo, we were able to see its urban planning strategy [in introducing the three axes]. The municipality of Torino must have had a clear strategic vision. Torino's citizens may have also participated in the planning. Citizens' participation is a typical approach to urban redevelopment.

In Japan, this public-citizen collaboration is rare. Especially until the second half of the 80s, the Tokyo Metropolitan Government took a strong initiative to redevelop Tokyo, such as in the development of the south center of Tokyo Bay. Yet, by the beginning of the 90s, the municipality of Tokyo lost initiative to developers and other private-sector entities. Developers led urban development in areas including Marunouchi, Roppongi, Shibuya, and Nihonbashi; the public authority became only a supporter for the development.

When I visited the office of the Port Authority of Boston, the then municipal architectural planner articulated the city's strategic vision. In contrast, Tokyo's public sector has no clear urban vision. In urban development, it is increasingly more important to have citizens' and public perspectives. In addition, links among small-scale initiatives – not necessarily large-scale – are becoming essential. Many architects in Japan lead in creating charming human-scale architecture.

Small-Scale Initiatives in Japan

Ginza has an interesting initiative. The building owners of Ginza formed an association, fought constructors, and won the 56-m limit on the building height. In Ginza, the town's commission examines all the projects in the area.

Yanaka has a tight-knit community built by locals. One major thread uniting the

ッパー）の関係性の再考が必要だと考えています。日本では、（3本の軸の導入を実施した）トリノやその他の米国都市同様に、自治体の強力なリーダーシップの発揮が必要です。2020年の東京オリンピック・パラリンピックのための再開発計画を見ると、国や自治体の足並みが揃っていないことがわかります。公的部門のリーダーシップ強化と同時に、東京の都市空間とコミュニティの個性を守っていくためには、地域に根づく草の根活動も進められるべきです。

community is a group led by Mrs. Mayumi Mori that continues to publish a local magazine. Through this and other efforts, locals' sense of security has increased in the community.

Relationships between Top-Down and Bottom-Up Approaches in Japan

My vision is to reconsider relationships in urban planning: relationships between the top-down and bottom-up approaches as well as relationships between the public authority and commercial developers in the top-down approach. In Japan, municipal authorities should take a greater initiative just like in Torino (in the case of the three axes) or some American cities. Japan's redevelopment project for the 2020 Olympic and Paralympic games suggests the lack of coordination in public authorities' initiative. At the same time, Japan should promote small-scale initiatives in neighborhoods to maintain characteristics of the urban space of Tokyo and its communities.

Participants

PoliTo	Marco Santangelo	Hosei University	Ayane Tateishi
	Claudia Cassatella		Chang Chia Hsin
	Nicola Russi		Chen Jiun An
	Mauro Volpiano		Chen Yue
			Gota Yamamoto
SCI-Arc	John N. Bohn		Hikari Okamoto
	Ramiro Diaz Granados		Ka U
	Casey Rehn		Karin Yoshihisa
	Hernan Diaz Alonso		Kazuki Maruyama
			Kei Ueta
Hosei University	Makoto Shin Watanabe		Masaki Oshiumi
	Koh Kitayama		Momoko Sakanashi
	Haruka Kuryu		Nguyen Quang Tuan
	Yusuke Shimada		Sakan Miyata
	Yutaro Muraji		Shinnosuke Tsubaki
	Masatoshi Hirai		Taihei Maruyama
			Takaya Shimodaira
	Hidenobu Jinnai		Takumi Anezaki
			Takumi Narimoto
Teaching Assistant	Ayaka Fujita		Wang Qingling
	Chisato Isome		Yo Housei
	Keito Kubota		Yo Ukyo
	Takeshi Osawa		Zang Lei
	Yugo Yasui		
	Yukiko Kondo		

Students

PoliTo	Alon Brozgal Shusterman		
	Giorgia Greco	The Editorial Committee	Makoto Shin Watanabe
	Lorenzo Fante		Koh Kitayama
	Monia Buongiorno		Haruka Kuryu
	Silvana De Bari		Aya Iida
	Valeria Vitulano		Wataru Date
SCI-Arc	Derek Yen	Translation	Rina Norikiyo
	Gabriela Zappi		
	Javier Benavides	Book Design	Tetsuya Mizuno (Watermark)
	Jianing Yang		
	Kai Wang	Photo Credit	TADA / YUKAI
	Liu Erdong		(pp.1, 18-19, 44-45, 80-81,
	Siying Chen		126)
	Shuchang Zhou		
	Slina Zhang	Editing	Speelplaats Co., Ltd.
	Soul Kim Min Kyu		(Jiro Iio,
	Ye Xiao		Mutsumi Nakamura,
	Yi-Wei Tim Chen		Daisuke Kawajiri)

江戸東京の都市組織に挑む

上野　本郷　谷中　根津　下谷

2019年9月30日　第1版 発行

編著者　法政大学江戸東京研究センター＋法政大学デザイン工学部建築学科＋
　　　　南カリフォルニア建築大学＋トリノ工科大学
発行者　下出雅徳
発行所　株式会社 彰国社
　　　　162-0067　東京都新宿区富久町8-21
　　　　電話　03-3359-3231（大代表）
　　　　振替口座　00160-2-173401
　　　　http://www.shokokusha.co.jp/
印刷・製本　株式会社 シナノパブリッシングプレス

©法政大学江戸東京研究センター＋法政大学デザイン工学部建築学科＋
南カリフォルニア建築大学＋トリノ工科大学　2019
ISBN987-4-395-32139-1　C3052

本書は2019年4月に電子書籍として刊行したものを新たに単行本として刊行しました。
本書の一部あるいは全部を、無断で複写（コピー）、複製、および磁気または光記録媒体等への入力を禁止します。
許諾については小社あてにご照会ください。